All the W

The hill seemed to go on for ever. At the bottom, it was overgrown and devious, forcing me to plant my feet carefully, searching for every hump and hollow in the grass. I pushed my way through a snarl of brambles, out into closely dotted bracken, gorse and bony-fingered elderflower trees. As the elderflower trees thinned, they became useful and I pulled myself up on any branch that looked as if it could take my weight.

There was no mistaking an elderflower tree. The flowers were smoky white and drowned everything in their cloying, acrid stench. I knew it well. Every summer, my dad brought home carrier bags full of flowers and hung them in the garden shed. Day by day, the scent grew stronger while the flowers frazzled and fell apart. Then he'd scrunch them onto sheets of newspaper, catching the dead flowers and scooping the final pile into a tin. Drunk daily, elderflower tea would, reckoned my dad, prevent colds and flu.

It must have been true. I couldn't remember him ever having a cold.

After climbing for a couple of minutes, I was high enough to see over the elderflower trees and even past the towering oak at the end of Mr Bowmer's garden. As I was wearing my green snorkel parka with the hood up, I hoped that neither Mr Bowmer nor anyone else would know it was me. The hill wasn't exactly out of bounds, but people in the row of council houses at the foot of the hill naturally assumed that anyone up there had nothing but mischief on their mind. Probably smashing something. Riding a motorbike over something. Stealing something.

I stopped and listened. Just my own breath, soft and muffled by the hood of my parka. Under my feet, the grass was dark and thick. Not like Mr Bowmer's lawn. This grass was tough and surly. You could grab hold of big clumps and tug at them, but they would always hold fast.

Sometimes, the grass would sneer at you. *'Who do you think you are? I've been here for a hundred years and I will be here for another hundred years. Go home now, before I catch hold of your shoelaces and pull you down-down-down underground. Under the trees. Under the rocks and coal seams. Under everything.'*

It's difficult to argue with grass.

Older boys often tried to start fires on the hill. At first, the flames would jiggle around like deranged puppets, but then a cold wind would rush up the valley and over the estate, quickly pinning the flames flat on the ground. Holding them down. Suffocating them. All that was left was a scorch mark, and that would be gone within a month. In its place was more grass, staring out at the houses and swishing with laughter.

It's even more difficult to argue with grass when it's laughing at you.

The last bit of the hill was much steeper and required an even greater concentration of effort. My scuffed shoes were splayed outwards now, feeling for any help that the ground might be kind enough to offer. To my left and right were a series of short, shallow trenches that I called 'bomb holes' . . . made, I always imagined, during lightning raids in the Second World War. Normally, I would have sat in the one on the far left and spent a minute or two watching the estate, but this wasn't a watching day. I walked straight past and onto the uneven table of waste ground that lay across the top of the hill. It was the size of two or three football pitches, blotted with hummocks, brambles and more gorse, and bordered at the back by another row of council houses.

The Undisputed
KING
OF SELSTON

The Undisputed
KING
OF SELSTON

DANNY SCOTT

JOHN MURRAY

First published in Great Britain in 2025 by John Murray (Publishers)

1

Copyright © Danny Scott 2025

The right of Danny Scott to be identified as the
Author of the Work has been asserted by him in accordance
with the Copyright, Designs and Patents Act 1988.

A CIP catalogue record for this title is available from the British Library

Hardback ISBN 978 1 399 81679 3
ebook ISBN 978 1 399 81681 6

Typeset in Bembo Std by Palimpsest Book Production Ltd, Falkirk, Stirlingshire

Printed and bound in Great Britain by Clays Ltd, Elcograf S.p.A.

John Murray policy is to use papers that are natural, renewable and
recyclable products and made from wood grown in sustainable forests.
The logging and manufacturing processes are expected to conform
to the environmental regulations of the country of origin.

Carmelite House
50 Victoria Embankment
London EC4Y 0DZ

www.johnmurraypress.co.uk

John Murray Press, part of Hodder & Stoughton Limited
An Hachette UK company

The authorised representative in the EEA is Hachette Ireland,
8 Castlecourt Centre, Dublin 15, D15 XTP3, Ireland (email: info@hbgi.ie)

For Hannah, Reg, Dad and Coal.
And Ria for the hours.
And Jim, Jocasta and Charlotte for making me see sense.

Somewhere near the middle, the ground was carved in two by a slender rift known as the Canyon. If it really was a canyon, it was a small one. Even at the widest point, it was no more than 20 feet; at its deepest point, no more than 30 feet. But its steep sides were covered in a soft, sandy earth that crumbled all too easily if you attempted to climb up or down. From the air, you could imagine a giant, jagged knife had cut through the hill, just for fun. The Canyon had no purpose. There was no river or stream running along the bottom. No piles of soil or boulders that hinted at a collapse in the years before the estate was built. It was simply a slice of the world that had gone missing.

Only now, standing at the top of the hill, at the top of the estate, at the top of the village, did I turn around. The wind was eagerly pushing and poking me and, as I looked out, I felt sure I could see each gust careening across the rooftops. Rattling trees and telly aerials. Pummelling drain-pipes and peevishly flapping any window that wasn't shut tight.

The sky was summer-storm grey and stretched menacingly in every direction: left, right, back over my head towards Mansfield and forward to the edge of the Peak District some 8 or 9 miles away. So far and so wide that its murky greyness took on an intense silver sheen. At this time of year, at this time of the evening, the landscape came to life. The never-ending wind would jostle and bully the dense clouds, leaving just a few holes for the westward-rolling sun. Only then could it reach out and touch the land, daubing everything with unruly splodges of cherry, tangerine and ginger. Bizarrely, the colours seemed to welcome the encroaching darkness, pulling it closer and closer towards the ground.

Straight ahead, there was the estate, then trees, then the little parade of shops and the school, then fields, then houses, then fields. All the way to somewhere else. To more hills and valleys and swards and pastures. To the towns and cities. On my left were farming fields, a pub, the Top Rec, houses, then another grassy slope that swooped up and down for a mile

3

or more. The main road in and out of the village followed its contours in a nervous, not quite straight line, then disappeared into trees that would have once been part of Sherwood Forest. In Robin Hood's day, the forest held sway over much of the East Midlands.

To the right was the church; built in 1120, it'd been there almost as long as Sherwood Forest. My friend, Collo, told me that the dark marks on the church door had been made by a witch in 1742. The witch's mother, egged on by gossiping villagers, had set fire to her only daughter, but she somehow managed to run from her house to the church, where she collapsed against the locked door. The poor girl's pounding, frightened fists had left charred patterns up and down the wrinkled wood.

When I asked the vicar about it, he said that it was black paint from when they replaced the hinges earlier that year.

There was another estate, nicer than mine, crouching in front of the church and a row of much larger detached houses beside it. This was where the schoolteachers, councillors and factory managers lived, and some of the houses had beautiful cars gleaming on their spotless driveways. After that, the land dipped and dropped and rose, then dipped and dropped and rose some more, before levelling out close to a line of red-brick factories, shiny offices and car parks for the managers and their beautiful cars when they weren't parked on their spotless driveways.

Way over to the left of the factories and offices was a tall chimney, surrounded by a collection of curious concrete buildings. To the right of the chimney, a vast, higgledy-piggledy metalwork sculpture was topped by two spoked metal wheels. My dad called it the 'headstock' and it was responsible for the safe movement of the lift that carried men up and down the hundreds-of-feet-deep shaft at the pit where he worked.

In front of the buildings, a procession of ordinary-sized lorries was constantly coming and going. To the far left was a shorter line of extraordinary-sized yellow lorries with gigantic knobbled tyres that churned across the mysterious grey mud of the Clay Heaps. This, too, was part of the pit, but it was worked by machines rather than men. One of the machines was called an excavator and I had often watched it glide across the Clay Heaps like an ancient, skeletal first-rate man-o'-war. Instead of a hundred cannons, it had a single spinning disc at the front which tore strips of grey flesh from the earth and carried them, via a series of belts and pulleys, to the waiting lorries.

Up on the hill, it was time. I pulled down my hood, shook my hands and stamped my feet. The red neon sign in the chip-shop window blinked on-off-on-off-on then settled for the night. The shop was almost hidden by the clouds' giant shadows, but I could just make out the shapes of two people inside, getting ready for the Friday rush. Next to the chip shop was another pub. It was close to seven o'clock and enthusiastic customers were already queuing outside, chatting easily and flibbing fags.

More lights clicked on. By the church, the top and bottom of the main road. Lights in front rooms, bedrooms and bathrooms. I unzipped my parka, all the way down. With each gust of wind, it curled and snapped, eager to let itself be pulled free from my shoulders and into the past. Slowly, I stretched out my arms and breathed in as much of the wind as I could hold.

As I tipped my body forward, the parka strained and yanked itself tight. I stumbled for a second, then dug my shoes deeper into the grass. Again, I tipped into the wind. The grass held me and held me, until . . . I rose a few feet from the ground, my legs dancing awkwardly as I struggled to steady myself. Up a few feet more. A few more. Suddenly, I was sailing across the houses. Above the chip shop and the bus stop. Up. Above the Co-op and the chemist. Up. Above the pit and the Clay Heaps. Up. Above the bobby-dazzlers and the fag-flibbers and all the other people I had known.

If I had to wake up, I hoped that I would wake up on another day. In another place.

A Deceit of Lapwings

The first thing I saw every morning as I pushed open the curtains of my bedroom window was Pye Hill Pit. It watched me as I dragged the glazed ceramic po from under my bed and went downstairs to empty the contents in the almost-outside lavvy. It nodded approvingly as I flushed the almost-outside lavvy and returned the glazed ceramic po to my bedroom. It was there when I played football on the lane at the side of our house and when I climbed the slide on the Rec up by the bus stop. It was still there when I tumbled off the slide and onto the grass. And last thing at night, as I knelt to use the glazed ceramic po before pulling my curtains back across the darkness of the bedroom window, all I could see was the pit.

During the long, warm, hazy summers of the early 1970s, I would sit on my bed and watch the flickering wheels of Pye Hill's headstock for hours. And on short, cold, clear winter evenings, the sun would set just to the right of those wheels, silhouetting them, the canteen, the men, the smoke and the heaps of coal against a hard, carefully polished metal-blue sky. As night-time blossomed, the lights in the pit buildings would send beams through the chilly air like golden arrows, hinting at the mechanical magic that was taking place inside and underground.

Who needed spaceships, jet planes and racing cars? Who needed Curly Wurlies, Weebles and quick-fit elasticated bow ties? I had the pit. And the best bit was my dad worked there.

In fact, Dad's job as a miner was the main reason we lived in that house. It was at the end of a small row of red-brick terraces known as the Pit Houses and they'd been built specifically for the families who worked at the local pits, brickworks and pipe yards.

Dad could walk from our back door to Pye Hill in less than ten minutes. And on Friday mornings in the school holidays, I would walk alongside him to collect his wages. Children weren't usually allowed on pit land. In cahoots with Pye Hill's magic and wonder, there was menace and risk: high-voltage motors and toxic chemicals, long threads of thick, steel wire that could easily tangle and confuse small feet, metal contraptions that might gash your face or crush over-inquisitive young minds.

Despite this ever-present danger, the offer of a walk to the main pit building with my dad was far too good to turn down. A rare chance to stand beside the men and machines that kept such a keen eye on me. A rare chance to marvel at what they did and how they did it.

I made that journey many times, but there is one that I remember more than the rest. It was spring. Quiet and not too cool. As soon as we walked through our garden gate and into the field, I knew that the world had planned something special for me and Dad. We were meant to be in that particular place at that particular moment on that particular day.

Dad was already ahead. Off duty, so his donkey jacket – the heavy, black Melton-wool coat worn by many National Coal Board (NCB) employees – had been swapped for his favourite green jerkin. Always unzipped, it flapped to and fro and to and fro again and again, keeping time with his footsteps. A pair of worn navy cords sat low on his hips and he was carefully rolling a fag as his cheap canvas trainers moved through the long grass at the edge of the field. He'd been smoking since he was fourteen and the creation of a fag was second nature, as effortless as combing his hair. All done on the move: Old Holborn tin out of left jerkin pocket with left hand, cracked open, one fag paper pulled from inside the tin and placed carefully between teeth, right hand now tugging at a tuft of tobacco, tin closed and put away with left hand which then retrieved fag paper, tobacco sprinkled along the length of fag paper, thumbs and first two fingers of both hands used to roll and shape the finished product, a dab of the tongue on the fag-paper glue, a final roll and shape, poked into the far right of his mouth, England's Glory matchbox out of the right jerkin pocket, match struck, cupped for protection, lifted up, a brief flare and . . . a slow, satisfying draw. Completed in less than thirty seconds, never once looking at his hands or breaking a single stride.

Just inside the long grass, the field was flecked with lapwing nests: twiggy clumps that blended beautifully with the cobbles of earth. A pair of lapwings watched us from a safe distance, and both let out a warning call as I got closer to one particular nest.

I stopped to look at the four eggs and bent down to inspect them.

'Tha mun' duh that!' my dad said sharply. 'They dunna like it if yuh touch 'em.'

I looked at the lapwings and, under my breath, reassured them that I would never touch their eggs. And if I ever saw anyone touching their eggs, I would chop their hands off. The lapwings seemed happy with this, so I saluted and they sealed the deal with another couple of friendly 'peewit's. One of them explained that they were more worried about badgers and foxes than me and Dad.

'Tha's raight,' I told them. 'Ah'll lukk after yuh.'

Thanking me, the birds took several steps closer. As they turned to the side, the light slashed across their bodies, revealing dazzling greens, bruised purples, blues and golds. Their dark, majestic crests sweeping up and back like high-tech, Brylcreemed hairdos. I had never seen anything so beautiful. So matter-of-factly immaculate in every detail.

I moved forward, desperate to ask why something so precious would choose to live amongst the dirt, the manure and the worms. Without warning, the lapwings were airborne, quickly followed by another pair to my right and another straight ahead. And many more. Climbing as one. A deceit of lapwings, wheeling left and over a line of small trees.

A group of lapwings is known as a deceit because they are supposed to be sneaky birds. Dad told me that. But he also added that, in his opinion, the lapwings had been hard done to.

'*They anna sneaky like people a' sneaky,*' he said.

Then he told me a story about leaving a penny underneath a lapwing nest one night as he was going to work. When he walked past the nest the following morning, the penny was still there, thus proving the lapwing's honesty.

While I bristled at the mighty injustice heaped upon the lapwings, Dad had carried on. I ran to catch up with his fag and his jerkin, and my running disturbed more birds. Crows, a yellowhammer, a pair of goldfinches, sparrows – and three or four rabbits playing on the freshly fertilised earth. All this, so close to our house. I couldn't think of anything else to do but wave both hands at the world and laugh. Waving my appreciation for choosing to share the day with me. Laughing at how happy it made me feel.

I was right. Today was going to be special.

The field came to an end at the pit border, but there were more living things – a rough-and-ready hawthorn hedge, false barley, sturdy battalions of rosebay willowherb, dande- lions, tall nettles and, holding a valiant front-line defence, the Meccano-like arms of an impressive cow parsley. I snapped off the thickest, longest length and asked Dad to make a blowpipe. He took a small penknife from his trouser pocket and sliced the ends, leaving a six-inch-long hollow green tube which he handed back to me. Without talking, we both headed for a patch of hawthorn hedge that proffered a few forgotten berries and gathered a handful each. I pushed a berry inside the tube and fired it across the mud, hitting a row of gas bottles.

Around our feet were tens, if not hundreds, of pieces of

thin metal. Straight-edged shapes that looked like straight-edged capital letters: E, I, H, L, T. As we got closer to the older, run-down pit buildings, I saw piles of these letters – discarded parts of old electrical transformers – shoved hurriedly into corners and doorways. And inside the run-down pit buildings were bits of oily engines strewn across oily floors and stacked on oily shelves. A chest of metal drawers appeared to be empty, but I found a few rolls of yellow wire in the bottom drawer. Like the lapwings, all these riches were camouflaged by their natural habitat.

And there, right in front of me, was Dad's natural habitat . . . Pye Hill Pit. One of the many pits in the Erewash Valley, birthplace of mining in medieval Nottinghamshire. Among Pye Hill's various shafts, No. 1 carried my dad to depths of 700 feet where he would work part of a vast coalfield – the best coal in the country, he reckoned – that ran all the way from Nottingham and Gedling, on through Hucknall, Kimberley, Eastwood, Brinsley, Underwood, Jacksdale and Selston, then north towards Pinxton, Kirkby and Mansfield. So incessant and generous were the coal seams beneath my feet that Pye Hill had been physically linked underground to other collieries at New Selston and Selston over a distance of almost 2 miles. A subterranean spider's web of not-so-secret tunnels and shafts, rail tracks and carts, cables and columns of men.

This was my home. And I was immensely proud of it. In return, my home carried itself well. Stout, unsophisticated and handsome. Calm farming fields that softened and zhuzhed up surly stretches of green and grey industrial meadow. Some steep hills and shallow valleys. Our skies skimmed with a harsh, heaving beauty that made my eyes glisten and my head tingle. And every inch of it coddled and attended to by coal. A concerned, paternal presence with reassuring sentry posts standing guard at every mile. Collieries and communities with names like Annesley, Bentinck, Pye Hill, Bull & Butcher, Selston, Moorgreen, seven shafts at the Portland site, three at Mexborough, plus the opencast quarries and spoil tips, engine sheds, tramways, gravel pits, chimneys, kilns, workshops, headstocks, reservoirs and wharves to allow coal traffic to access the Cromford Canal.

Strewn in between, like second-hand glitter, were the people. Us. The dirt-encrusted pubs, shops and homes of the unassuming alchemists – thousands upon thousands of men, women and children – who had magically turned black gold into gold. Families in Lincolnshire's agricultural heartlands grew potatoes and cabbages. In the West Midlands, they worked metal and made cars. In the six towns of Staffordshire, they fired clay. And here, in rural Nottinghamshire, we worked coal. Lots and lots of coal.

All of this was my home and I was happy here.

As Dad approached the gates of Pye Hill, he turned and waited for me. We were now surrounded by lorries and men. Lots of men. Some in overalls, some in their own raggedy shirts and worn-out trousers; each man obscured by dirt, dust, sweat and exhaustion. Each man balancing a dirty white helmet jauntily upon his dirty head. Most of them had moustaches and all of them dragged their heavy boots with heavy legs and heavy feet. They joked and swore. They smoked and coughed and laughed and coughed.

And coughed.

After picking up his wages from a small office with a small window, Dad headed towards a pair of doors that opened onto a wet floor. The room was lit by a long line of fluorescent lights, but the billowing steam made it difficult to see where I was. Naked men dried themselves with large towels, while some washed under heavy jets of water. I listened to the various noises that pinged around the walls: more coughing, more shouts, more laughter, some singing and the clang of metal lockers, all carried along on the relentless shoosh of gushing water.

Although this was the place where men washed themselves, it was covered in dirt. The same black dirt that I saw under Dad's fingernails or nestling in the corner of his eye. In his trouser pockets and on his lips. Ground forever into the nape of his neck and the many scars on his back. Trails of it slid like thin snakes from the shower stalls to the drainage trough, but every time one disappeared, two or three more would take its place.

Dad walked past the showers, nodding and occasionally speaking to men in various states of cleanliness. He stopped by the row of metal lockers that made the clanging noises. Opening the end locker, he pulled out a long black plastic case and slung the strap over his shoulder.

Even I could tell that it was a rifle. And I naturally assumed that Dad was going home to shoot my mam.

The Ronco Chop-o-Matic

Mam and Dad had been married for a long time before I was born and they were older than most of the other Pit Houses parents; Dad was mid-forties and Mam early forties when I arrived. They must have enjoyed each other's company at some point because I had seen black-and-white photos – never on show, always tucked away in tins and sideboard drawers – that proved they had once laughed and drunk and sat around with friends. But something must have happened in those earlier years. If there ever was love, it was now a gristly, suspicious kind of love. A bruised kind of love. Occasionally given a hefty clout. The kind of love that too many women in Selston expressed via black eyes and split lips on Monday morning as they went to the shops or caught the bus to Alfreton.

Although I never saw my dad hit my mam, I heard it. Only once or twice, but it did happen. And she hit him, too – that I did see. Wild, haphazard thumping – delivered with a mix of grunts and angry words – across Dad's shoulders and face. It would all start over something simple like Mam cooking his Sunday dinner at twelve o'clock when she knew he wouldn't be back from the pub till two o'clock.

Mam's approach in the kitchen – whether it was chips, broth or Sunday dinner – was one of intense, niggardly defiance. I was never asked to help and knew better than to offer. Me offering help would have suggested that I didn't think she was capable of cooking Sunday dinner on her own. And that would come at a cost.

'*Cumm 'ere, ah'll gi' yuh bluddy 'elp.*'

So, if the weather allowed, I disappeared up to the Rec or spent Sunday morning booting my ball against the side wall of the house. If the weather didn't allow, I sat in the front room and read my comics and books. Or cut out interesting bits from the paper. A picture of soldiers on the streets of Belfast or Fred Earwaker, director of finance for Quaker Oats Limited, standing in front of some boxes of Sugar Puffs. An advert for Timex watches: 'It's a gift at £3.10'.

Occasionally – very, very occasionally – I would secretly stand in the passage and watch Mam's kitchen conflict unfold. Helpless carrots, potatoes and turnips yanked from the pantry, steadfastly despatched as if she was slaughtering chickens. Saucepans and knives clanging around the range and kitchen sink like church bells blasted free from the belfry. Those poor vegetables boiled for an eternity, like Christians in ancient Rome. A beef joint roasted for an eternity like . . . well, also like Christians in ancient Rome.

The battle would always cease just before twelve and the result was always the same: total and complete victory for my mam. The vegetables, having bubbled for an hour or two, were now bloated and lifeless. The beef joint, having sizzled and screamed for three or four hours, was hacked into rough hunks. Everything then strewn across three plates.

'*Weer a' yuh? Cumm an' sit dahn!*'

That was the signal for me to stop whatever I was doing and sit at the kitchen table. I ate as I assumed all small children ate: whatever was on the plate ended up in my stomach. Quietly. No fuss. Even when I didn't know what was actually on the plate. Mam ate with the same level of wrathful vengeance that she cooked with: stab-stab-stab-cut-cut-cut-chomp-chomp-chomp.

Dad's dinner sat steaming on the table in front of a third empty chair. I watched it while I was eating my dinner and Mam finished hers. I watched it until the steam was exhausted. Every week, Mam expected Dad would arrive in time to eat with us. Every week, she was disappointed. Mam and me could have eaten a bit later, but we didn't. So she would shove Dad's plate of food into a still cozy oven, where it would slowly cook for another couple of hours until he came home from the pub.

Monday to Saturday, Dad worked nights, leaving home around eight or nine in the evening and returning at eight or nine in the morning, depending on how much overtime he'd done. But Sunday was different. Having returned at eight or nine in the morning from the Saturday night shift, he'd immediately go to bed, sleep for three or four hours, dress in his smart shirt and trousers, then head for the Dog & Doublet, the pub at the bottom of Jubilee Hill, the hill that bordered the bottom of our row of houses. Often accompanied by two or three other miners who were also enjoying their two or three hours of freedom on the Lord's Day.

Miners were well known for their ale intake, but generally speaking Dad didn't overdo it. Just a couple of pints, then back up the hill – often accompanied by those same miners – and into our kitchen. Nothing was said. He simply wrapped a tea towel around his hand and plonked the still-scalding plate on the table. Eating his now shrunken, brittle dinner was a lengthy affair and could easily take half an hour or more, with every crunchy mouthful separated by a couple of minutes of snoring. Sometimes, I would sit next to him, dipping a teaspoon into a pot of jam for my pudding and wittering on about the Timex advert or the man from Quaker Oats Limited. Eventually, my mam would bustle into the kitchen, snatching his plate and whatever food was left. Roused and growling, my dad would spring into action.

'Ah anna finished yit!'

'Yuh shudd 'a cumm homm before, then,' answered my mam, slamming the plate onto the table.

And that was it. More shouting by my mam. Dad pointing his finger and Mam reaching for something to hit him with or throw at him. A mug. The salt cellar. A rolling pin. The Ronco Chop-o-Matic.

My mam's voice. *'Ah'll bluddy wang yuh. Talkin' t' may like that.'*

Dad snapping that he was, *'Norr allowed t' eat me dinner in peace.'*

Getting too close to the action was dangerous, so I'd stand at the passage door, watching them stamp and gesture at each other in the dimly lit kitchen. Surrounded by the black iron range, deep sink, old-fashioned utensils and colourless walls, they looked like two obstinate ghosts, despatching well-worn words and phrases from a time long before the Ronco Chop-o-Matic had shredded its first cabbage leaf.

Young children naturally think that mams and dads are designed to fit together like Lego bricks; manufactured for each other's benefit. But growing up with my mam and dad left me puzzled. They never . . . talked. Forget the shouting, I mean talking about regular stuff. Other mams and dads did it. I heard it when I was over the road at Clive Willoughby's house or up the road at Dickie Boom's house. Those mams and dads talked about a new shop that had opened in Alfreton or that squeaky cupboard door by the window. They might even sit around the kitchen table together and laugh at the same things on the radio.

Although my mam and dad lived together, that was about it. They were two people with very little in common, press-ganged not onto a ship bound for the Cape, but into a life bound for . . . nowhere, really. Just drifting about for a bit before drifting back to where they started.

Their mismatch was physical, too. Dad was no Engelbert Humperdinck, but, on the whole, he was presentable. Working class, yes, but clean-shaven and kempt. A regular haircut. To the casual observer, an approachable man, if a tad subdued. My mam, on the other hand, was approached with caution. More than a tad outraged and often on the warpath. Carelessly dressed . . . even for a miner's wife. At five years old, all I had for reference were the other mams that lived in the Pit Houses, but compared with those other mams, Mam came across as the poor relation. Her clothes had holes in them and stains on them. Her ridiculously frizzy hair was rarely combed and stuck out at uneven angles. She had filthy hands on blotchy arms and equally filthy knees on filthy legs that were given an even angrier, scabbier, swollen look by ignored and untreated cellulitis.

Mam was squat, like a tiny, tubby darts player or a tiny, tubby boxer. Thickset, not quite 5 foot. A suitably tough miner's wife for a suitably tough miner like my dad, living in a suitably tough mining village like Selston. Anyone doubting that toughness need only have glanced at those hands. Small, quick and covered with working skin that was as firm, rough and ridged as a good-quality cheese grater. Her nails were tiny, but it wasn't because she chewed them. They'd just been beaten into submission by years of washing dishes and windows; sweeping yards and bringing in buckets of coal for the fire; scraping wet laundry across a washboard in the depths of winter. At some point, the nails must have realised it was best if they kept out of the way.

Sometimes, those maltreated nails were spruced up . . . painted with red nail varnish. Half a dozen times a year, for a wedding or a Saturday night at the Dog & Doublet. A rare enough occurrence for a young child to notice.

The sad thing was, even when the nails were coated in red nail varnish, they still looked lost and abandoned.

Mam smoked a lot and drank a lot. As well as the Sunday afternoon scuffles with my dad, she had proper fights. With women *and* men. Once, at the Portland Arms in Jacksdale, I noticed a commotion by the bar and saw my mam punch a man in a hat. And I heard her yelling 'Yo can bugger off!' as he was sent sprawling into a row of chairs. I saw her push a woman off our back yard step when the woman started shouting at my mam . . . something to do with money. She, too, was sent sprawling: this time into the copper – a gas-operated, free-standing boiler that heated water on wash day – which gave her a nasty gash on the forehead.

Part of me was delighted and proud. In the way that children are. My mam can smash your mam's face in. In fact, my mam can probably smash your dad's face in as well. My mam can carry heavy wooden furniture up and down the stairs. My mam can clear a foot of snow from our front yard. With bare legs. In her slippers. And did my mam ever ask for help? Why would she do that? Mam wasn't like those women who lived in the large detached houses up by the church. The wishy-washy wives of the wishy-washy school-teachers, councillors and factory managers, with their beautiful cars gleaming on their spotless driveways. Wishy-washy wives needed help. My mam didn't.

For a while, the delight and pride at having Henry Cooper for a mother gave me a sense of well-being. Don't you dare touch me or I'll tell my mam. But then I realised that Mam's cut-and-dried idea of law and order applied to me just as much as it did the man in the hat and the woman on the back yard step. Perhaps I failed to appear at the kitchen table bang on the stroke of twelve one Sunday afternoon because I was enjoying an extra-long spin on the Witch's hat at the Rec. Or I foolishly asked my mam what the woman on our back yard step had been shouting about. Was I given a wild thump, a punch or pushed into the copper? Of course not. I was a small, defenceless child. With a terrible haircut and shorts that wouldn't fit me properly for another three or four years.

For me, there was . . . the Strap. (Ta-daa!) A leather strap with a metal ring at one end that was designed to clip over a bathroom tap. The other end had a sort of handle which allowed you to pull the strap tight and use it to sharpen cut-throat razors. Like you see in pictures of Victorian barber-shops.

Was Dad aware of the Strap? (Ta-daa!) If he was, I don't remember him pulling me aside to offer any comforting words. I don't remember him wrenching the Strap (Ta-daa!) out of Mam's hands.

Dad was aware of the house fires, though. House fires in our house. Not serious fires. Just a scorched armchair or dress, resulting in first-degree burns to the face or legs. Mam was a heavy smoker, even for the 1970s. Three to four packs of twenty every day; sixty to eighty fags that needed sixty to eighty matches brought into close contact with my mam's comically curly, sticky-out hair. Whether distracted by something on the radio or simply rushing to light another fag, she could easily find herself a couple of inches off-centre with the match, causing one side of her hair to burst into crackling life.

It happened several times a year. Sensing that she was on fire, Mam would let out an *'Ooh-er!'* then start slapping her head and face, letting the match fall onto the armchair/ tablecloth/carpet/skirt. Considering the flammability of 1970s fashion and furnishings, self-immolation was a distinct possibility. Thankfully, I was in the front room with her on the one morning when the match really did manage to get its fiery teeth into the armchair and Mam's skirt. The skirt blackened and melted as Mam leapt to her feet, but the flaming tassels on the armchair were starting to look busy.

Although I knew that you were supposed to put water on fires, I was too small to reach the taps. I had seen Clive Willoughby's dad put out a chip-pan fire by smothering it with a towel, so I grabbed a tea towel from the kitchen and started madly padding down the flames. It worked.

Mam didn't seem that bothered. She just threw away the skirt, covered the scorched armchair with a sheet, opened all the windows and told my dad that it was my fault. She insisted I'd been *'Bonnin' stuff in t' grate'*, but he looked at me . . . and I looked at him. He knew.

Now you understand why I assumed Dad might want to shoot her.

The View from the Top of the Bus

Everyone can remember their first day at school. I can even remember that I was wearing grey shorts – still far too large – and red sandals. As I sat on the bus from the Pit Houses to Selston Church of England Infant School, I kept looking down, past the grey shorts, at my red sandals, wondering if they were clean enough for school. Would I be made to take them off if they were too dirty?

My dad was with me, which meant he'd either taken a holiday or finished earlier than usual. We sat upstairs, of course, so Dad could quickly roll and quickly smoke a quick fag in the six or seven minutes it took to get to school. Riding on the bus made me feel important, like I was going to work. Most of the other people on the top deck were smoking, chatting, reading newspapers or having a go at all three. Visibility was severely hampered by the Player's No. 6 pea-souper, but you could just about see through the windows if you shoved your face right up to the glass. The top deck of a bus was a rare chance to view my world with fresh eyes. Up there, you could see the church tower, which instantly added a sense of distance to the journey. Ahead and to the right were the houses that had once belonged to the Bull & Butcher Pit. On the left, farming fields and fences that gradually ebbed away towards Pinxton and Somercotes.

At the far edge of the last field, I noticed one of those strange and solitary clumps of trees that the farmer hadn't bothered to move. He'd ploughed and planted around it. Left it as it was, like a treasure island. It was about the size of a small house and I thought how wonderful it would be to live in that House of Trees on my own treasure island. To sit in the narrow band of long grass that surrounded the tree trunks and eat my breakfast in the sunshine. To listen to the crows' indolent caws or the far-off honk and rattle of the 8.30 train from Alfreton to Nottingham. To be able to look out and see the blue buses rumbling by.

That was the moment when I learned to fly. Nudging past Dad, I climbed out of the back window of the bus and onto the roof, where I used wind power, the padded 'wings' of my open anorak and the speed of the bus to launch myself up and over the church, across the graveyard, all the way to the solitary House of Trees. From there, looking back at the road, I could see myself on the top deck, fourth window from the front: the slightly worried face of a small boy who was going to school for the first time. But no one on the bus could see me. Not even my dad. Not even the other me.

On the bus, that other me was almost at school and Dad was getting himself sorted. Old Holborn tin back in his pocket; a single matchstick to clean out his ears and loosen any lingering flecks of coal dust that were stuck behind his nails. And finally, his brown-with-gold-bits comb, swept once or twice through his grey hair.

'A' tha ready, lad?' he asked.

Yes, I think I was.

So many children. One or two faces I knew, but the rest were strangers. More children than I'd ever seen before. Could it be that every child in the world was going to my school? Thousands of them . . . millions of them. All filing through the one door, filling the playground with movement and noise and cautious glances. So many that Dad had to pick me up and carry me into this place. He said a few words to a pointy-nosed woman who pointed to another door, then he carried me through that door and talked to a not-pointy-nosed woman in a mahogany-brown, heavy, woollen, chequered two-piece suit. She pointed to a chair by the open fire. He lowered me onto the chair and talked to the second woman again.

I was pretty sure I'd never seen my dad talk to any other women apart from my mam and Vera, who lived next door to us. What did Dad and this school woman talk about? What did he tell her about me? That I was good at reading? That I liked gardening? And cutting out pictures from the newspaper? And listening to music on the radio?

To prove my dad wasn't lying, I stood up and bawled the opening lines to 'In the Summertime' by Mungo Jerry, a recent number one hit. The woman smiled, but Dad looked uneasy. Maybe he hadn't heard.

This time, I would stand on the chair and . . . too late. Dad was already gone.

I pushed the chair closer to the fire and sat down, watching the hot coals and listening to the scattered sobs and wails of the nervously excited children. Dickie Boom's son, Richard – also known as Dickie Boom – was one of them. Easy to spot with his tightly curled ice-white hair. We sat next to each other for the rest of the day and drew things. One girl, Stella Curtis, had drawn children skiing down a snowy hillside. It was all done in felt tip and packed with comforting details like bobble hats and snow on windowsills. Sledging I could have understood, but skiing? How did she know about skiing? How did she know what people looked like when they were skiing?

I don't think my dad came with me on my second day at school. Or third. Or many of the other days. I do know that most of my journeys to school were made on my own or with other Pit Houses kids. Sometimes there was an adult – never my mam – sometimes not. Sometimes I would carry a comic. It was called *The Topper* and it was the same size as a tabloid newspaper. I'd seen men reading newspapers on the bus and thought that reading *The Topper* would help me blend in. I noticed that the men sometimes shook their heads and muttered at their newspapers. So I perfected my own shake-shake-mutter-mutter. They also closed and folded their newspapers in a hurry when they were getting near their stop. No problem. Halfway up the hill, past the Methodist chapel, rustle-rustle-fold-fold, into my anorak pocket.

There were mornings when the bus never arrived. Severe weather in the winter, industrial unrest, occasional break-downs. And then? I walked. Sometimes with others, sometimes on my own. Along the first half-mile of that walk to school, there were no houses and not much traffic. The road had a gentle lift-dip-lift with a slight curve, like a vast grey wave, frozen in mid-cycle. In the dip, you couldn't see anything ahead or behind, just the gradual rise of the tarmac. And no one could see you.

Selston had one oldish man who was rumoured to be a kiddy fiddler. He lived not far from the school and had been seen talking to some young lads on the Bottom Rec. One of the lads told his dad a lurid tale of trousers at half-mast and bags of sweeties. Full to the brim with understandable rage, the lad's dad went to the oldish man's house and kicked his cat to death, then hung it on the washing line.

I remember the teachers talking about it in class, looking very serious and telling us to stay away from the oldish man's house. Then, a few days later, the lad admitted that he'd made it up. The oldish man had been talking to them, but all he'd done was tell them off for wazzing in the sandpit. One of his neighbours got the oldish man a new kitten, but the poor fella died not long after it'd all happened. The kitten went to live with the neighbour.

Although there wasn't much bullying at the infants' school, there were a couple of lads you needed to look out for. One of them lived right at the top of the row of Pit Houses, but he lived in a non-pit house. That, according to my mam, meant they had more money than us. He certainly acted like he'd got more money than me, regularly remarking on the poor quality of my jumper and highlighting the unevenness of my haircut. I have not forgotten his name, but let's just call him . . . Sadistic-Lad-Who-Lived-Next-To-The-Bus-Stop-In-A-Non-Pit-House.

He had this stock routine where he'd shout my name, all friendly like – *'Ay-upp, Danny!'* – then, once I was within arm's reach, he'd pinch the skin on the side of my neck, twisting it while he looked at me with his twisted face and laughed his twisted laughter. Then he'd make me sit on the floor and rest his knees on my shoulders, gradually transferring his weight until my scruffy little body was bent double like a safety pin.

He used to nick my copy of *The Topper*, too.

Despite these setbacks, I liked school. A lot. I liked being asked to do things. I liked being able to do those things. Within a few weeks of starting at Selston Church of England Infant School, it became pretty clear that I was a Clever Bugger. Spelling and adding up, reading and writing, colouring in, drawing tanks and aeroplanes . . . Danny Scott was a small, snotty-nosed Renaissance Man to rival the best that fifteenth-century Italy had to offer. I rattled off facts and figures about the triceratops (courtesy of something I saw in the newspaper). I made a poster that explained the water cycle (copied from an encyclopaedia in the big classroom). From the encyclopaedia, I also learned the capital cities of far-off countries like Chile and Laos (just before it changed its name to the Lao People's Democratic Republic). I was even the best singer in class, belting out 'O Jesus, I Have Promised' with a surprisingly lusty dollop of religious zealotry.

Be thou forever near me? My master and my friend? Hallelujah! Praise the Lord! Now, let's have you up on the cross. And don't spare the nails.

Although Sadistic-Bus-Stop wasn't in my class, someone had obviously tipped him off about me being a Clever Bugger. In amongst his twisted laughter, I could sometimes hear him whispering, *'Grotty Scotty! Non s' clever naa, a'yuh?'*

Even at five and six years old, these words made no sense to me. If anything, his neck-skin-twisting laughter only increased my 'cleverness' because I started learning things just to spite him. The correct spelling of pterodactyl. How to sharpen Dad's lawnmower blades. How to dance to Dave and Ansell Collins' infectious 1971 reggae hit, 'Double Barrel'. The more he twisted, the more I learned. A virtuous – although painful – circle of knowledge.

I even won prizes; prizes that I knew would piss him off. Extra gold stars, sets of pencils and books.

My mam seemed equally pissed off by the prizes.

'Ah left school at fourteen an' went in t' service. Ah canna read 'n' write, burr it nivver did may any 'arm,' she said when I brought home the book I'd won for being a Clever Bugger at spelling. *'Wot d' yuh want t' be like that for?'* she asked, genuinely confused by what was happening to her son. *'Answerin' questions an' showin' off all o' time.'*

Should I have tried to explain that this wasn't something I had any control over? It was just happening to me, like getting taller or losing my milk teeth.

My bedroom was tiny, so I decided to take my collection of Clever Bugger Prizes to the House of Trees. Then it really would be a treasure island. First, I needed something to put them in. Something sturdy, yet easy to hide. Perfect! An old green first-aid tin, about the size of a Weetabix box. More than enough room for two books, a Tufty Club badge, one of Stella Curtis' ski-slope pictures, a scuffed Anadin tin full of sixpence pieces and an unopened Milkybar.

My flying adventures were a regular occurrence by this point and, that night, I placed the green first aid tin next to my bed, naturally assuming it would travel with me as I whooshed over to my usual hangout, the House of Trees.

In my musty pyjamas, I climbed out of the bedroom window and onto the roof of the Pit House. My bare feet scrambled easily up the tiles towards the chimney pot and, from there, I could see the church tower. Somewhere between me and it was the island and my House of Trees.

I ran down the roof and jumped. As usual, I struggled with height and balance at first, bobbing around like a lost balloon, but everything started to settle as I thrust my head and arms forward, straining and stretching them toward the church tower. Always making sure the first-aid tin was safely stashed inside my pyjama top.

Despite the early hour, the countryside below me was brightened by a uniform yellow glow that came from the main pit building. Gentle enough to be easy on the eye, but strong enough to highlight the tiniest detail on the far-off land that dashed below my rippling pyjamas. It took no more than a minute to locate the House of Trees.

What I heard was not the sound of night. It was the lumbering putter of a tractor and the incessant hum of summer insects. It was the crows and lapwings, sparrows and blackbirds. It was the lazy rustle of leaves and the slow-motion wafting of branches. The taller, broader trunks of the House of Trees lined the outside in a rough rectangle, closing together at the top to make a magnificent roof of chameleonic greens. Inside, there was more than enough space for the various branches and logs that had toppled and twirled themselves into surprisingly comfortable items of furniture – bouncy chairs, sturdy benches and even a clothes rail. Sunlight shone through the green roof, liberally colouring me, my pyjamas, various flowers and berries, the bits of furniture and everything else inside my secret home. Even the air was green. Of course! This was the ideal place to hide my stash. How would anyone find a green first-aid tin in a green world?

I slowly slid it underneath a dense and honest-looking holly bush, then sat down on one of the sturdy log benches. Watching the buses and looking out for small boys going to school. Listening to the birds and the trains. Still no sound of night. Except underground. There, I heard the night. The sound of machines and men. Always the machines and the men. The mangling grind and grab of the longwall trepanner. The bash and clatter of the coal carts.

Back at the Pit House, my bedroom was neither green nor filled with the sound of machines. It was still tiny and smelled of fag smoke, but . . . there was no first-aid tin. My treasure was safe.

My regular flights, though enjoyable and well established, were not guaranteed. Take-off didn't always work when I wanted it to or when I needed it to. There were times when I was desperate to rise above whatever was happening on the ground, desperate to feel the cool air on my face and marvel at the miles of criss-crossed hedges and fields, but . . . nothing. I would leap from fences, climb the highest trees, run and howl across the Clay Heaps. Still nothing.

Why did it let me down on that day at the Rezzer?

At the right side of the field at the back of our house was a grubby square of water that had something to do with the pit. About 30 or 40 yards by 30 or 40 yards, coal-black, still and, for the most part, silent. Known by everyone as the Rezzer. The water was only 4 or 5 feet deep, but below the surface there was another world: a dense jungle of old mattresses, unwanted pets, cupboard doors and car parts. Some of them occasionally poked into the daylight – a potato sack filled with kittens. Some of them could be used as climbing frames if you swam closer to the middle of the water – oil drums and petrol tanks. Some of them had gone under and stayed under – broken windows, coat stands and metal things with sharp hooks. These hooks were cunningly hidden by grasping, unbreakable weeds, but always eager to catch a flailing arm or slice open a leg.

None of this stopped me or anyone else going there on summer afternoons, when the Rezzer was treated like the local lido. Afternoons that were thick and bloated with sunshine. So dense that you could feel the heat parting as your body squeezed through the air. Each step taking longer than the last.

On that day, my slow steps along the Rezzer bank were stopped dead by the laughter of two boys who sounded much older than me. Old enough to smoke. Old enough to swear. Crouching by the hedge, I heard them loudly effin' and jeffin' as they walked up the track from the main road. I spied through the hedge, decided I didn't know them, and thought it best if I rolled silently into even deeper grass. They were fourteen or fifteen years old. One of them wore a brownish T-shirt that was too small for him; the other had a light green jumper tied around his waist, revealing a faded red football shirt. As they stripped down to their underwear, they continued to laugh, smoke and swear, occasionally flicking fag ends and burning matches into the water. In between the laughing, smoking and swearing, they shot quivering mouthfuls of phlegm in high, long loops that flopped down on the Rezzer surface with a rich and crispy slap.

I decided to wait until they were both in the water, then I would make a dash for the stile that led to the field that led to our house. Football Shirt took a short run and jumped into the water. Brown T-shirt sat down and lit another fag. I listened to a minute or two's swimming and smoking, then silence. Then more silence. Brown T-shirt shouted something but continued to smoke. Another shout or two and the fag was finished. And he was in the water.

I counted to thirty before starting my run but then saw Brown T-shirt struggling to swim with something heavy. At the sloping sides of the Rezzer, thick slime made it difficult to climb out, so he began scooping with his free arm, trying desperately to move towards the shallow corner where I was now crouched. Brown T-shirt pulled Football Shirt out of the water and I could see he was bleeding heavily from his leg. Brown T-shirt was making strange noises; Football Shirt was silent and still. Brown T-shirt ran towards the main road in his underpants. Y-fronts, like most boys wore, but being older than me his were coloured. Sickly purple. Finally, I started for the stile, but couldn't help looking down as I passed Football Shirt. His eyes were closed, his wet skin covered in goosebumps, left leg oozing blood from the thigh.

White Y-fronts. Why was he still wearing white Y-fronts, at his age? White, but turning red.

I wanted no part of this. No association. Not a witness who could swear to this or that. Not a passer-by. Not here, not now. Not someone who was involved. Not someone who happened to be there. Just by being there, was I guilty? I was pretty sure that lying in the long grass was no excuse.

So, instead of climbing the stile, I tried to fly. In my head, I flew, but I wasn't flying and I knew I wasn't flying. I jumped and closed my eyes and wished and wished some more. I jumped hundreds of feet in the air, so high that I had to wear a space helmet like the one my mate, Graham, got for his birthday. And yet there I was, on my belly. Crawling like a lizard or a wanted man, arms and legs pushing and pulling. Eyes closed, I saw nothing. Eyes open, I saw almost nothing. Just the green of grass and weed. Never was I so happy to see grass. And a stile. Oh, stile, my resolute and generous friend.

Up and over, running beside the tall hedge that marked the bottom of the gardens in our row of houses. Too close to the hedge; skin nicked and picked apart by hawthorn and bramble. I felt but couldn't see. I had SEEN nothing. And I SAID nothing. When I stopped and climbed a small gate at the bottom of the garden two doors down from us, I still said nothing. There was a shed, close to the gate and close to the hedge. Just enough space for a small boy to squeeze in between. A space to lie. And tell lies. I saw nothing and waited.

It wasn't easy. Every tremor of a twig, every bark of a dog and vroom of a car. The heat joined in, inflating itself and rolling out like a paddling pool, squashing down and against my side. I was going to get the blame for this. I knew it.

Countless stories turned over in my mind, but none that offered a concrete alibi. So I stayed there till I heard an ambulance and some shouting. I stayed there till the ambulance went and the shouting stopped. I stayed there until it started getting cold. Still, I said nothing. I went home. And still, I said nothing. The walk home felt like the end of a long journey. Tired and thirsty, like I'd walked from Crich Stand on the Derbyshire horizon.

In the kitchen, Dad was making his 'snap', the sandwiches he would take to work that evening. Mam was in the front room, smoking.

Dad looked at me. *'A' tha raight?'*

I nodded and watched him make his 'snap'. An artist at work. Be it peanut butter, pilchards or dripping, each was applied as a master plasterer attending to the best wall in the best house in the whole of Nottinghamshire. The knife dancing across the bread in swift, almost careless strokes. Almost careless. A dab here, a dab there. Swish-swish. Not until every corner and curve had been given a peanut butter/pilchard/dripping makeover was the slice folded and laid to rest in the 'snap' tin.

Dad knew I was watching him and raised himself from the chair, ready for his other kitchen-table trick. A cup of tea. No milk, always black. The same plain mug for as long as I could remember. Starting with the teapot at the mug's edge, he would tip the spout in small increments, anticipating the moment of tea-fall. And then, as the mug gradually filled, he would raise the teapot higher and higher until his arm was above his head, and then standing on his chair, arm touching the kitchen ceiling. His aim never faltering; the tea always entering dead centre with loud splatters that built into a fine head of dark brown bubbles.

More than once, I shouted, *'Dad, it's bubblin' ovver.'* But it never did.

'Weer yo bin? Did yo ha' owt t' doo wi' that at t' Rezzer?' Mam's voice brought me back. Where had I been?

'Upp at t' Rec.'

And did I have anything to do with that at the Rezzer? Silence.

And then: *'A few on us were upp at t' Rec. There's a blackbod's nest in t' edge.'*

30

That extra detail clinched it. Everyone was satisfied that I had been up at the Rec. Even I believed myself.

Although the sun continued to sit heavy on the world for several days, the Rezzer was off limits. I looked and listened from my bedroom window but neither heard nor saw any of the usual shouts and splashes. As I looked and listened, I thought about death. Had I seen death before? Just the kittens in the Rezzer. And the rats and birds in the bottom of hedgerows. Did that count?

For a few nights after that day, I dreamt about Football Shirt. He was always wearing a space helmet and kept complaining that he couldn't breathe. I wasn't wearing a space helmet, but I couldn't breathe either. Then my mam was there, sitting at the kitchen table, but she didn't have a head. Her head was on the floor and the head kept telling me she was dead. In the dream, I looked out of the kitchen window at the sky and saw two moons, one large enough to touch. I reached up and thought for a moment that my reaching had lifted me from the ground.

Then I would wake with a jerk. Still in my bed. Just the one large moon glowing outside my window, rinsing and wiping everything clean. Lighting the garden path and the hedge at the bottom and the field beyond the hedge.

Finally, I realised what I had to do.

It took me less than five minutes to get downstairs, out into the garden, through the gate, across the field, over the stile and into the long grass next to the Rezzer. The grass where I'd seen Football Shirt was still flattened, guarded at the sides by much straighter, sterner-looking grass. I climbed into the hole and positioned my body like the body I remembered. Such a powerful moon. Bright white light. On my way to heaven . . . I shall not be moved. No wonder I was so good at singing hymns.

It was warm enough to sleep and I slept until I sat up and pulled my damp pyjamas away from sticky skin. Two lapwings perched on the fence in front of a fiery blue morning sky. I didn't have a watch, but I sniffed and looked at the colours in the sky, working out the hour from where the brightest yellows were and how much of the sun I could see.

Would my mam notice I wasn't in my bedroom? I decided to go home, just in case.

Invisible Things

Our house – the Pit House – came with its own kind of darkness. I'm not talking about the set-tos and the parental disputes. It was physically dark. As if it existed in tandem with the coalface at Pye Hill. The rooms were cramped, the windows smaller than they should have been. Each room did have a single electric ceiling light, but they were rarely switched on. It got to the point where I didn't even notice the ever-present gloom and became convinced that living without electric light had given me super-eyesight, like a fox.

That was the nickname I gave myself for much of 1971: The Fox. I would hone my super-eyesight in the front garden, reading car number plates before they reached the bus stop at the top end of our row of houses. Long after Mam had gone to bed, I would sneak downstairs to the kitchen and make cheese and cream crackers or a Nesquik milkshake in total darkness. And with complete silence. Beware of The Fox! He breaks into your house and makes himself a snack.

After The Fox, I became The Shadow. Quietly slipping through the night like . . . well, like a shadow. Superpowers? Similar to The Fox, really. Without the snacks.

On my own, I enjoyed the dark house; it gave me the chance to experience a flavour of Dad's world, 700 feet underground. I enjoyed it less when the 'relatives' came round. There was an uncle who wasn't really an uncle. A young man, mid-twenties, with spiteful grey eyes that matched his spiteful grey smile. He only appeared when my dad wasn't there and would take food from the pantry or money from my mam's purse. Every few months, he would smash open our coin-operated gas meter and walk off with a bulging pocketful of shillings.

Sometimes, he would bring a woman and the woman would bring a bottle of something. The man, the woman and my mam would sit in the kitchen, drinking whatever was in the bottle. Then they'd all start singing and clapping and dancing around the table. While the man was singing and clapping and dancing, he would pick up tins from the various shelves, poking around inside to see if he could find any money or jewellery or watches. He didn't try to hide his actions or pretend that he wasn't poking around inside tins. Still, my mam said nothing. So I said nothing.

After the bottle was finished, my mam would stagger upstairs, laughing – giving the man a chance to poke around the sideboard in the front room – only to reappear wearing bits of my dad's St John Ambulance Brigade uniform. Dad was a corporal and his heavy black uniform was always hung on a special wooden coat hanger at the top of the stairs. It had two silvery-white stripes on each arm, just below shoulder patches that said: 'St John Nottinghamshire'. The peaked cap had a white band, decorated at the front with a silver metal St John badge. In the white plastic shoulder bag, he kept smelling salts, bandages, safety pins and his ceremonial white cotton gloves – worn for the regular parades.

The Association was formed in 1877, teaching first aid to ordinary working people. Volunteers were taught to splint broken limbs and stem bleeding wounds, and a close relationship soon developed between the brigade and the mines. When serious injuries happened deep underground, men like my dad were first on the scene. Men like my dad saved lives.

Dad never said anything to me about the St John

Ambulance Brigade, but I knew that it mattered to him. It mattered more than most things. It mattered enough to have its own special wooden coat hanger at the top of the stairs. It mattered enough for my mam to get a barbarous thrill from taking the piss after she'd necked a few sherries. The man and the woman would fill the kitchen with renewed crinkly laughter and song as my mam came through the door, squeezed into a swimming costume, over which was Dad's St John jacket, the ceremonial white cotton gloves and peaked cap.

'*Cumm 'ere, ah'll wrap a bandage rahnd it,*' she'd cackle as she gyrated suggestively past the Ronco Chop-o-Matic. '*Lukk at 'is bluddy silly 'at,*' she added, waving the peaked cap. Suggestively.

There were other visiting relatives and other types of madness. An aunt from Alfreton who wasn't really an aunt, always accompanied by her daughter, my cousin who wasn't really a cousin. The cousin would kick things off, waving her arms and singing strange songs about ponies while the aunt stood in the corner of the front room, face turned towards the wall, rocking aggressively on the heels of her clumpy shoes. The aunt would also hum and mumble to herself, which gave an odd counter-rhythm to the cheerful pony songs.

Although I wasn't unduly worried by the strange behaviour, I fully expected my mam to say . . . something. To at least ask the aunt what she was up to.

But all my mam said was, '*D'yuh want a cupp o' tea?*'

And if nobody answered, she'd ask another question.

'*Which bus did y' catch?*'

At some point, the aunt would take things up a notch. '*Ah'm gooin' t' duh me-senn in!*' she'd howl. '*Ah've 'ad enuff.*'

This further roused the cousin and she'd begin skipping around the sofa, still singing the pony songs.

'*My-boyfriend-is-a-pony-and-ponies-don't-tell-lies. Green-grow-the-rushes-O.*'

Noise, movement, madness . . . exploding in a surreal

crescendo as the aunt strode towards the fireplace, removed the fireguard and plunged her foot into the blazing grate.

Silence no longer seemed my best option. *'Mam! Lukk wot shay's doin'. Shay's moved t' guard an' purr 'er futt in t' fire!'*

Thankfully, the aunt pulled it out sharpish when her thick tights began to melt, but then switched to banging her head hard against the wall, each thud timed with her rocking and mumbling.

'Mam, shay's bangin' 'er 'ead on t' wall!'

She only managed a few thuds before she fell backwards, whimpering like a baby.

'Mam, shay's fell ovver! Shay's scraitin'!'

Even as this woman cowered and shook on our front room floor, my mam seemed not the least bit surprised. She didn't . . . notice. Just swivelled her head to different bits of the room, trying to make polite conversation. *'Did y' know Chuffin' Billy's wife's gone in t' ospital?'*

I soon began to notice there were quite a few things that Mam didn't . . . notice. She never noticed if a light was on or off. She never noticed what I was wearing. If I kept quiet when she was in the kitchen, she never noticed I was there. Was that so strange? My mam was busy, chuntering to herself, washing pots and smoking. No wonder she never noticed me.

But my interest was piqued and, a couple of days later, I was in the front room, watching Mam wipe down the windows. It was something I'd watched many times but, this time, I watched closely. She missed several bits and didn't even see Vera, the lady next door, waving at her. Were the windows invisible? And Vera?

So I began to experiment, creeping into rooms, looking at Mam to see if she was looking at me. I rattled drawers and cupboard doors, waiting for her eyes to suddenly shoot in the direction of the rattled drawer or cupboard door. She would stand still and silent, like a wolf in the forest, listening for the clumsy tread of humans. Yes, she was looking at me . . . but she didn't see me.

I started to make a list of things that were invisible to my mam: me, Vera, muck, chairs, thieves and madness, cups and light. And smashed glass.

Once, while she was upstairs, I threw a glass onto the kitchen floor and waited.

'*Wot a' yuh upp tew?*' she shouted.

'*A glass fell off o' table.*'

'*Ah'll gi' yo' glass!*'

As she stomped downstairs and into the kitchen, I positioned myself prominently by the window, pointing at the glass on the floor. As quiet and still as an old, tattered photo. She saw neither me nor the shattered fragments and walked towards them.

'*Wot a' yuh upp tew?*' she repeated.

Foolishly, I answered. '*Ah towd yuh. A glass fell off o' table.*'

Again, the wolf's keen hearing pinpointed her prey and she marched over the broken glass towards me.

'*Ah'll gi' yo' glass!*' Her arm whirled through the air like she was still wiping down the windows or slashing at enemy soldiers on a samurai battlefield. I was close enough to look into her eyes now and saw they . . . were . . . e . . . m . . . p . . . t . . . y. The more I looked, the deeper the emptiness. The deeper I sank, the more I looked. Transfixed, I forgot to move and the whirling arm caught hold of my sleeve.

'*Ah'll gi' yo' glass!*'

As she raised and flung down her other arm, I tugged my sleeve free and ducked. She missed. Several other swipes rained and ranged but, like a nimble boxer, I ducked and shimmied, rolling under and around, then circling out to the right and behind her. She continued to swing and growl but I was now by the back door, out and round the side of the house to the kitchen window, where I continued to watch her arms wave and cuff the air.

My mam was blind.

No wonder she never reacted to the grey, thievish smile. Or the attempted suicides. No wonder we never switched on the lights at night. No wonder she set fire to her hair. No wonder she always answered the door with '*Ay-upp. A' tha raight?'* even when she didn't know who the person was. No wonder she never wrote a shopping list or darned my dad's work socks. No wonder she never read the paper or looked at the letters that tumbled through the letter box. No wonder she never helped me get dressed in the morning or came with me on the school bus. No wonder she was a bit rubbish at making tea.

Not long after I found out about my mam being blind, a woman who lived up near the Rec, Mrs Oates, brought over a small child, Rebecca, in a huge, springy pram and asked my mam if she could watch the child for the afternoon.

'Course ah can.'

Mrs Oates explained a few things, left a couple of nappies and a feeding bottle on the table, and went out to catch the Alfreton bus. I sat down – quietly – in the corner of the kitchen and watched. Mam seemed to have an idea where the pram was and shuffled towards it. She reached out, trying her best to be careful, but still shoving her arm into Rebecca's face. Immediately, she started crying and the pram began to rock on its springs.

'Can yuh see a bottle?' my mam shouted, but I didn't answer.

She swept both hands across the kitchen table and knocked the bottle onto the floor. Now on her hands and knees, she groped forward, left and right. She was nowhere near it, so I used my foot to roll it towards her hands. It caught her wrist, causing the spring-loaded fingers to quickly scoop it up. Somehow, she managed to prop the bottle into Rebecca's mouth and lean it against the side of the pram, then sat down for a fag.

I wondered if the match would fall into the pram. I wondered if the milk would choke Rebecca. I wondered how she was going to change the nappy. Or see if Rebecca had been sick. I watched as she smoked several fags and the baby gurgled. I watched as she started to rock the pram with far too much enthusiasm, making the bottle fall and the baby cry again.

Mam soon grew tired of her new toy. She stood up and

moved the pram so she could get past. It bumped into the guard that surrounded the open flames of the cooking range, the hot colours reflecting in the pram's metal frame, springs, handle and spokes. In the front room, I heard the radio and the strike of another match, so I moved the pram back a few inches, then a few inches more. Inside the pram, I looked for Rebecca but all I could see was me. How had my mam managed? How had she changed *my* nappy? What had happened if *I* was sick or *I* tried to climb out of the pram? What had happened if *I* got too close to the fire?

From that moment, I started to really-really-really watch. To become a man of science and not merely a petty experimentalist. To note and understand Mam's every move. To work out how she lived.

This is what I learned. My mam spent a lot of each day cleaning, but the house was never clean. Stains were never tackled unless I or my dad tackled them. Dropped carrots that rolled into corners would stay dropped and in corners for weeks unless I or my dad or Vera found them. Outside the house, Mam seemed to know – more or less – where she was on the back yard, but rarely ventured to the front yard on her own. On the rare occasions when she did, it involved an ostentatious sweeping of the path or clearing of snow. She never popped over to a neighbour's house to borrow a packet of tea; she always sent me. Her cooking was hit-and-miss. Her appearance was hit-and-miss and often involved mismatched slippers. Genuine emergencies when my dad wasn't there – a badly cut finger – were signalled by a continuous thump on the wall that divided our house from Vera's house. Vera would appear and Vera would know what to do.

Luckily, our out-of-the-way home spared Mam much regular contact with the rest of the world. Our needs were largely catered for by men in vans. The Milk Man (and his van), Bread Man, Corona Pop Man (fizzy drinks), Butcher (chops and mince served from the back of a blue Bedford CA), Grocery Man, Fish 'n' Chip Man (a van with frying apparatus in the back), Paper Man (newsagent), Betterware Man (household goods, brushes, butter dishes and the like) and Clubby (a woman who would collect weekly payments for whatever had been ordered from the Grattan catalogue: 'The Lanter ORLON/NYLON Blanket. Soft, light and warm. Drip-dry, mothproof and stain resistant. Bound at each end with nylon ribbon. 20 weeks @ 40p'). Each of these people would appear like clockwork on their allotted day, delivering their respective goods. Bills were settled, with monies entered into a thin column on a coloured card.

The allotted days and times were important because it meant that my mam knew who to expect and when to expect them. If it was Monday at 11 a.m., Alan the Insurance Man. If it was Friday at 5 p.m., Todge the Grocery Man. Like an overexcited audience of one, I bought tickets for every performance of *Knock On The Door*, silently applauding Mam's inventive recital.

It's Friday . . . it's five o'clock. *'A' tha raight, Todge? Eh, them carrots last wikk were luvvly. A' yuh seen Vera? A' they finished cuttin' 'edges upp by t' church? D' yuh know wot time A'f'ton Market closes on Sat'dee?'*

Todge the Grocery Man might tell her things. *'They've closed t' road off upp top o' Barrows Hill.'*

'Ah know,' lied my mam.

'They 'ad summ bad floodin' dahn in Pye Bridge, din't they?' said Todge the Grocery Man.

'Ah know,' lied my mam. Of course she knew.

'There's bin a noocleer explosion next t' t' Institute in Somercotes. Thahsands dead,' said Todge the Grocery Man.

'Ah know,' lied my mam. *'Ah were upp theer when it 'appened. Made ever such a racket.'*

She was there. She saw it all happen. How could she be blind if she saw it all happen?

For the things that couldn't be delivered or banged on the wall for or lied about, there was always me. I was sent to pick vegetables from the garden and sent up ladders to clean bedroom windows. I was sent to borrow money from neighbours and return said money. I was told to gather clothes pegs and washing that had been dropped on the back yard.

For those emergencies that were even worse than a badly cut finger – Mam running out of fags – I was sent to a tiny corner shop, at the bottom of Jubilee Hill in Pye Bridge. Over the River Erewash, under the railway bridge and past the scrapyard. At first – from the age of four or five – I walked. Going to the shop was easy, all downhill, but back home was a trial for tiny legs. Sometimes, a bus driver would toot, stop and open his doors for me, dropping me off outside my house without me even having to tell him where I lived. Sometimes, lorry drivers did this, too. And Todge the Grocery Man.

The journey was made easier after the arrival of my RSW 11. My slightly rusty but always trusty blue Raleigh RSW 11. I first spotted it leaning against a fence inside the scrapyard on the way to collect fags. Chain hanging off, back tyre flat, odd pedals and one handlebar grip missing. I stared at it for a few minutes before running over to the shop and the all-important Player's No. 6 brown packet. Then I crossed back over the road and stared at it for another minute or so. A man in a trilby, white shirt, greenish trousers and black braces walked towards me. I panicked and headed for the bridge.

'*S' tha want it?*' he shouted with a nodded smile.

I said nothing. Always the safest option.

'*Ah dunna want nowt forrit.*' He smiled again.

The RSW 11 was wheeled towards his office, turned onto its back, the chain wangled onto the sprocket, slormed with oil and a long air hose brought the back tyre to life. He tested the brakes, made a couple of adjustments and finally wheeled it towards me.

'*S' tha Norm's lad?*' he asked.

I wasn't sure what he meant.

'*S' tha live on t' corner? Upp top?*'

I nodded but, again, said nothing.

He leant the bike on my side of the fence and walked off.

'*Can ah 'ay it?*' I wondered out loud.

'*It's thine, Danny-lad.*' He knew my name. It must be mine.

I tried pedalling but quickly remembered that I had no idea how to ride a bike. So I pushed it home, gave Mam her fags and hid the bike in our garden. Over the next week or two, I spent hours practising on the grass at the side of the house. Falling more than riding, I soon attracted the attention of other children, including Sadistic-Bus-Stop.

When I finally felt ready for the road, I steered the bike up our front path and onto the causeway. Sadistic-Bus-Stop was quick off the mark, walking towards me. I guessed what was going to happen and waited till he'd covered half the distance. Wait . . . wait. Like a flash, onto the scuffed white seat, pedalling like the clappers, weaving like a drunken snake, off the causeway, into the middle of the road and over the brow of Jubilee Hill. And down. Wind and flies against my face. A blur of hedges and wildflowers unfurling at my side. My plimsolls now dragging along on either side of the bike to help me keep my balance.

By the time I was at the bottom, I could no longer measure my speed. Hundreds of miles per hour. Thousands. A car going up the hill honked its horn as I passed through the short tunnel underneath the railway bridge and the noise began to dart about the walls like an out-of-control firework. Sharp, honked pings that whizzed over my head and through the spokes of my RSW II.

Laughing and breathless, I appeared from the other side of the tunnel, pulling on my brakes and forcing my plimsolls harder into the tarmac. Except there was no tarmac. My feet were in mid-air. And below me, there was no Pye Bridge. My RSW II was sailing over a lush cartoon-coloured valley. The same lush cartoon-coloured valley that was depicted on the label of a jar of strawberry jam we had in the pantry. Rolling hillsides of luxuriant bottle-green grasses and comic-strip trees that had been designed and drawn by a masterly hand. Friendly rabbits lolloping across a perfect horizon with a perfect sky and chalky whiffs of perfect clouds.

I saw flowers that were much bigger than the flowers in Pye Bridge, shaped like liquorice wheels and Anglo Bubblys. I saw a main road that was nothing but a muddy track, bordered by a rickety, old-fashioned fence that everybody liked to sit on in the ever-present sunshine and moonlight. Its only vehicles were covered wagons and horse-drawn buggies, but beside me, up amongst the sunshine and moonlight, dazzling spaceships whizzed across a dark-but-alight sky.

Like most six-year-olds, I had little cause to reason why. Sometimes, you simply accepted that life was like that. You were given a bike by a man in the scrapyard and you disappeared into the label on a pot of jam.

As I waved to the spaceships and shouted happy hellos to the horse-drawn buggies, I repeated a simple mantra: *Don't forget the fags.*

Knowing that my mam couldn't see gave me a certain sense of independence. Freedom from scrutiny. As well as making my own way to school, I started pedalling up to Church Lane on my trusty RSW 11, visiting my friend Collo – like me, Collo was at Church of England Infants; he also, like me, owned a copy of *The Hamlyn Guide to Birds of Britain and Europe*, which was more than enough to forge an instant and seemingly unbreakable friendship – on Saturday mornings and staying there for the whole day. At seven and eight, I was an experienced traveller; more than capable of catching a bus to Alfreton Market and the accompanying high street of shops.

I perhaps took the 'freedom from scrutiny' thing a bit too far. Like the man with the grey eyes, I began stealing money from the tins in the kitchen. Just a few pence, here and there. Then, I started stealing from other houses, like Collo's and Duggy Dugg's and Dickie Boom's. Again, just a few pence. Nothing that would be noticed. I stole from the shops in Alfreton: sweets and chocolate from Woolworths, toys and trinkets from Vernon Hill, clothes and utensils from the Army & Navy Stores. I made my own breakfast and washed my own clothes. And I still watched my mam. I didn't let on that I knew about her eyes, I merely probed and prodded.

But this was no longer just about my mam. It concerned all of us – me, Mam and Dad – and how we lived and how long we were going to survive before the house burned down. Did my dad know my mam was blind? Did he know that I knew? What would happen if he died? Who would look after us? Who would look after me?

Like most miners, Dad didn't say much about . . . anything. Yes, he might argue with my mam when he'd had a couple of pints but most other times he was silent and rather shy. Unobtrusive. Invisible, like Vera and the windows. His silence only broken if someone had left the back door open.

'Y' lettin' all 'eat aht!'

In the name of science and progress, I tried asking him a few questions about my mam. About us. About Mam being blind. About what might happen. About stuff I'd learned at school. About the world and why it was the way it was. But I never got very far.

Not that I minded. Dad and me got on. We would walk down to the pit to get his wages or go swimming in the Rezzer and we'd talk about lapwings and the best time to pick blackberries. We'd plant carrots or mend the fence in the garden and talk about his Black & Decker D500 electric drill or whether the compost heap needed turning over. That was more than enough. I was happy with my lapwings and the D500.

Unfortunately, I was getting to the age when I also needed some real-world support. Somebody to fill in a form or accompany me on a school trip; somebody to watch me play football or give me a hand making a scrapbook for my homework. But the more real-world stuff I needed from

Dad, the less he had to offer. The main problem was that he always worked nights and was asleep in the day, so I hardly ever saw him. No chance of him filling in that form or accompanying me on a school trip. One of the first things I learned to write neatly was Dad's name, address and simple signature so that I could fill in my own forms. I was fond of my dad and I think he liked me, but it's hard to feel close to someone you know so little about. Even when it's your dad.

And my mam? Although she'd loomed large in my short life, I knew less about her than I did my dad. We didn't even have walks to the pit or swims in the Rezzer to get the conversation going. No lapwings or D500s. Granted, there was the blindness and the anger. I also knew – because she told me two or three times a week – that she'd *left school at fourteen an' gone in t' service*. I knew she couldn't read or write. She had a mother who was dead and half-sisters I'd never seen. She liked smoking and she occasionally set fire to stuff, probably because she was blind. I suspected that blindness fuelled much of her anger, but I also suspected that she was just . . . angry. Generally. About everything. If I asked her a question about lapwings or making a scrapbook, she became angry. She would sneer and laugh.

'Yuh think y' better than me, don't yuh. Yuh think yuh know it all. Ah were in service at fourteen!' If there was a visitor or someone she could put on a bit of a show for, she might add, *'Lukk at 'im, readin' 'is books. Ay's non s' clever when ay's wettin' bed every naight. Ooh, it duzz bluddy smell in 'is bedroom.'*

I mentioned this to one of my teachers and a nice lady from school came to the house late one afternoon to talk about how I was getting on and whether everything was all right at home. Dad was in bed and Mam wasn't sure what this posh lady wanted, so she just kept asking if I'd been in trouble.

'Way've 'ad t' call t' police, y'know,' Mam explained. *'When ay brokk next door's winnduh. Ah were in service at fourteen!'*

Next door's window had been broken, but not by me.

And the police were never called. How would she have got up to the phone box? Or had she got me to go up there and grass myself to the coppers? Mam liked telling that story, though, just so she could say, *'Way've 'ad t' call t' police, y' know.'*

I have no idea if the nice lady got the information she needed, but she came and sat with me the following morning and said that I was to go and see her if there were ever any problems at home. Strictly speaking, I should have told her what happened after she left, but I didn't. Mam opened the sideboard door and pulled out . . . the Strap. (Ta-daa!)

Knowing she was blind, I was sometimes lucky enough to spot the Strap (Ta-daa!) danger early and managed to make a run for it. Sometimes, my nimble boxer act – duck and shimmy, roll under and around – would get me out of trouble. But it's harder to duck and shimmy or roll under and around a flailing leather strap with a metal ring at one end. And when it does hit you, it hurts. So you stop dead and shout, which immediately gives away your coordinates.

Someone once remarked on the thick red lines across my back and legs. My mam chuckled and proudly called them 'Sergeant's Stripes'.

It sounds harsh, but . . . it wasn't. What I mean is that, although the Strap (Ta-daa!) hurt, it didn't feel 'over the top'. Other parents – even the ordinary ones that talked about cupboard doors – had similar accoutrements: a stick, a cane, a shoe, a police truncheon, a home-made cat-o'-nine-tails. The madness of life in the Pit Houses was actually a strange kind of normal and I really don't think I had it any worse than most of the other kids. In some ways, I might even have been slightly better off. Other kids had fathers who kept regular hours and mothers who could see, and both were on hand to deliver clouts round the ear.

But my dad was always in bed and my mam was blind. I could eat chunks of cheese without fear of reprimand. I could wee out of my bedroom window at night and wear Dad's St John Ambulance hat in bed. I could dance around in the back yard and draw tattoos on my arms. I could steal money and go shopping. I could safely bring home all the things I bought or stole. I could use the stolen money to order things out of the newspaper and, provided I intercepted the postman, no one would know.

My 'Sergeant's Stripes' were a small price to pay for so many glorious freedoms.

The Men in the Shed

There was a man in my dad's bed who didn't look like my dad. He was sitting up, resting his head on Dad's pillow and reading Dad's paper. I could see Dad's hair, his tattoos and the shape of his arms, but this man was covered in bandages and bits of gauze. Dull brown. The colour of old tea. The face looked painfully sore and swollen, and it had a beard, which Dad never had.

I saw tussled blankets and the heavy, crochet-squared counterpane. I saw dust in the air, meandering through the shafts of sunlight. I saw Dad's Old Holborn tin on a chair. But this man couldn't be my dad. For a start, he wasn't asleep; this man was sitting up in bed, gently poking his bandages. I stood by the door, looking at the face I didn't recognise and listening to my mam downstairs, clattering around the kitchen.

Dad had dermatitis. Not regular, run-of-the-mill dermatitis. This was irregular, not-messing-about dermatitis. In between the bandages and gauze, I saw scraps of glistening flesh, all yellow, red and runny. When Dad moved, the movements were slow and stiff. He tried to scratch his scabby nose, but bending his arm lifted the gauze and crinkled the sores near his elbow. The nose remained unscratched, and he gave up with a few hard-blown breaths.

My mam wasn't keen on 'official' people like doctors coming to the house. She was convinced they used to gossip about us . . . tell people that we'd only got lino on the upstairs landing and had holes in our tea towels. So when the doctor came to see Dad a couple of times, I knew it was serious.

Dermatitis was a constant worry for face workers – the men who went underground and gathered the coal from a coal seam – like my dad. Sometimes, they were helped by machines; sometimes, it involved hands and hand tools. At some pits, they would descend to depths of more than a thousand feet, in temperatures that matched a glorious, sunny summer's day. Sometimes, they would have to work knee-deep in water that had forced its way through the black walls. It was salty and full of minerals, and it would sting bare skin. The heat meant most face workers were semi-naked; if the skin had been nicked or grazed by lumps of falling coal or rock, that sting would become a burn, and the burn would be irritated further by the abrasive coal dust.

Once back on the surface, the men would attempt to scrub their skin clean, which the skin wasn't always pleased about. Even at home, asleep in bed, miners would continue to scratch at their forearms, scalps, ears and knees. Some of them would eventually get dermatitis. And if the dermatitis was severe, it stopped them going down the pit to chop away at the coal seam.

Despite his many shortcomings, Dad had never before given me cause to wonder. His shortcomings, like everything else in his life, were consistent and indefatigable. Along with Alan the Insurance Man and Todge the Grocery Man, Dad's life ran like clockwork. But this time the clock had been overwound.

There was no point in bothering Mam with any of my thoughts and worries about shortcomings and clocks, so I decided to go and see the Men in the Shed.

Pye Hill, along with several other pits in our bit of the East Midlands, came with an area called the Clay Heaps. As well as deep-shaft underground mining, there was a separate opencast section of the pit. Literally digging coal from the surface – or just below the surface – of the grey, clay-heavy earth that made up much of Selston. The clay that had been dug up needed to be stored somewhere, and that place was just to the left of the pit as you stood at the back of our house. Undulating grey dunes that baked hard and cracked open in the summer months, like staring at the surface of the moon. In winter or heavy rain, though, they acquired a topcoat of thick sludge that was impossible to walk on.

Several men worked on the Clay Heaps, driving diggers, dumper trucks and lorries, and their office was a large shed not far from the side of our garden. I was a regular visitor and the Men in the Shed's manager never seemed to mind me being there.

'Y' raight, lad?' he would ask, if he found me warming myself by the Calor Gas fire. 'Wokkin' 'ard?'

Sometimes, I did work on the Clay Heaps. I made cups of tea, cleaned muddy boots and checked the tyres on the dumper trucks. A couple of months before Dad's dermatitis episode, I lost my footing as I was walking over to the area where the trucks were parked. I fell backwards, then began to slide forwards, carried by the clay slime and picking up speed as the ridge got steeper.

'Les!' I called out to a man who was putting derv into one of the trucks. He saw me, just as I flipped over a ledge and down the side of a conical pool of viscid grey water. It was only 3 or 4 feet deep, but I was unable to stand on the greasy bottom and kept falling back under the surface. I could feel the cold in my ears and eyes, running up my nose and down my throat. It tasted chalky and bitter.

Suddenly, I felt a large hand grab my arm and pull me upright. I tried to wipe my eyes and open them, but they were glued together. I was hauled out of the water and could sense urgent running, accompanied by loud, concerned shouts. Inside the shed, I was placed by the Calor Gas fire while several towels and rags wiped me down. I opened my eyes and looked at Les. He laughed. And all the men laughed. Not snidey laughing. They smiled at me and patted me on the shoulder.

'Way've all dunn it,' said Les. 'Tha's one o' us naa.'

Another man wrapped me in a large coat and sat me on

the manager's chair by the desk. I was then given a mug of sugary tea and all the biscuits I could eat.

The Men in the Shed would definitely know what was going to happen to my dad and his dermatitis.

As I opened the door, the shed felt lively and was full of smoky warmth. The men continued talking about work for a few seconds, then one of them asked, *"Ow's y' dad?'*

I explained my concerns.

"Ay'd better git back t' wokk,' laughed another. *'Ay still owes may ten bob.'*

Les, the man who'd rescued me from the clay pool, was eating a sandwich. In between mouthfuls of white bread, he told me he was taking one of the lorries down to the pit. He wanted to know if I fancied a run out. My dad and Les had been friends for a long time. They went swimming together in the Rezzer.

Les drove up to the main road where Vera's dog had been run over. Everybody thought Sandy – Vera's dog – had a terrible accident, but her death was my doing. It happened during my brief 'magician period'. To be honest, the 'magician period' was little more than me finding a bit of wood shaped like a magic wand and trying to make things vanish. At first, I tried it on my mam and Sadistic-Bus-Stop without success, but then started waving the wand at trees and cars and . . . dogs. Sometimes, I would attempt my magic tricks while relaxing, listening to the radio or kicking a football at the side wall of our house. I'll be honest, I wasn't fully concentrating on the football and one shot skewed off the wall and sent the ball into the road. Sandy had been watching with interest and, being a dog, decided to chase it.

I thought the bus had missed her. She didn't bark or anything. For a second, she looked like she was going to make it across, but then she realised something was wrong. Her back legs had gone all limp and she looked back at me. She looked unhappy and a bit confused. Then she dragged herself round in a little circle and lay down about a foot from the kerb.

Had Sandy technically 'vanished'? Was it due to my 'magical powers'? Although there were arguments for and against, I can't imagine the Magic Circle would have been that interested.

Anyway, Les let a couple of cars go past the spot were Sandy had 'vanished', then turned left and headed towards Jubilee Hill. I had walked and ridden my RSW 11 down that road to Pye Bridge – and into the label from a pot of jam – many times, but riding up high in the lorry, everything looked different. Like riding on the top deck of the bus, I was being offered a glance at the world with fresh eyes. I could see over the hedges and over the trees, all the way to the Peak District. To immense green hills that wrought the horizon into an up-and-down jaggedy line. A formidable, hard-wearing horizon.

To the right of the road, the fields levelled out and a small flock of sheep were idly munching their way towards a farm at the far end. Below them was a dense line of trees that marked the edge of the River Erewash and I quickly traced them further and further into the distance, wondering where they'd stop. They didn't.

On the left was a rolling, fat meadow that spilled knapweed, poppy and corn chamomile all the way down the hill, through the hedge and almost into the road. The sun must have been just where it needed to be, lighting up not only the meadow, but also the swarms of bees, butterflies and buzzing insects that seemed to vibrate in the air above the purples, reds, whites and yellows.

Swallows were zigzagging across patches of the meadow, gulping down six-legged treats, while starlings darted among the flowers. A line of sparrows was perched on the hedge, watching the action unfold, oblivious to the old man sitting on the bench halfway up the hill. Les pipped his horn and waved at the old man, who didn't wave back.

'Ay's fast-on.' Les smiled.

Even with the pit buildings, gravel mounds and headstocks dotted across so much of Selston; even with the pipe yard, the coal dust and the Clay Heaps; even with the grubby stone railway bridge that crossed both the road and the river at the bottom of the hill . . . I felt at peace in this landscape. Its rough hills and trees and unfussy villages suited me. It was its own kind of pretty.

I looked at my dirty knees and picked at the snot trails on the sleeves of my jumper. I thought about the old man asleep on the bench, his trousers tied around his waist with a piece of white twine. Some people may have turned their nose up at those things but they made no difference to me or the houses or the fields or the swallows or the river.

At the bottom of the steep hill, Les took a sharp left onto a single-track road that followed the railway line all the way to the pit. In fact, the railway line ended at the pit and was used to transport coal in long lines of metal wagons. I had never been all the way down this road, past the various painted signal boxes and cabins that had been set along the railway line. I was pleased to see that, even here, amongst the rock, steel and concrete, there were flowers, weeds and shrubs that had refused to give way. Quietly, but defiantly, they were reminding everyone that they were here first.

'Is y' mam oaraight?' Les asked.

I looked across at him and immediately understood that he was asking a different question. This is the one I heard: 'What happened?'

After the incident where I fell into the water at the Clay Heaps, Les had walked me back to the house and stood halfway down the garden while I pushed open the back door.

'Mam! Ah fell ovver in t' Clay 'eaps!'

"Ow did yuh bluddy manage that?' She came bustling through from the kitchen, guiding herself with one hand on the wall.

'Ah slipped ovver.'

'A' y' clothes dotty?'

'Got summ mud on 'em.'

She grabbed my crusty, tangled hair.

'Yo goona get such a wang.' As usual, her free hand started flapping wildly, hoping to hit an arm or earhole.

'*Ay slipped when ay were 'elpin' us.*' Les was now standing behind me. '*A' tha raight, Hilda? It's Les.*'

Startled, she looked towards the voice and laughed, nervously.

'*Way gen 'im a wipe dahn,*' Les explained. I turned to look at him and saw he was tapping the side of his nose. I wasn't sure what it meant, but it made me feel better. '*Ay just needs a gudd wash,*' he told my mam. '*Ay'll be raight.*'

Looking at him from the passenger seat in the lorry, I decided to answer both questions.

'*Mam's oaraight. After yuh left, ah just 'ad me tea. An' a bath.*'

Les screwed up his face and turned it back to the road. He scratched his chin and shook his head. Then he shook his head and scratched his chin. Interfering in other people's affairs was tolerated up to a point, but what a man or woman wanted to do in their own house was their own business. How they treated their family was their own business. And in a place like Selston, neither Les nor anyone else could change that. Did Les ever hit his wife or slash his son's legs with a leather strap? Probably. But it was none of my business.

Nothing more was said as we trundled towards the pit, but Les reached out his left hand and pulled me across to the driving seat. He sat me between his knees and placed my hands on the steering wheel of the lorry. I don't know if I was really 'steering' but that made no difference. My hands moved and the lorry moved, turning left and right. Obviously, I pipped the horn a few times, frightening a couple of wood pigeons, and tried the indicators. But it was the 'steering' that stayed with me. It felt strangely satisfying . . . being able to direct this behemoth. Preventing the lorry from going out of control and smashing into someone's house or bowling over the train tracks.

Why didn't people come with a steering wheel?

Pulling into the main pit entrance, we aimed the lorry at the stores and stopped in front of a large ramp. Les climbed down and walked around the front of the lorry to open my door. It wasn't easy with short legs, but I managed to reach the metal step and jump onto the concrete.

He handed me a coin. '*Goo an' get y'senn summat frumm t' canteen.*'

I ran around the corner to the canteen block and pushed through the doors. The noise stopped me dead. A volley of garbled voices and laughter, all set to a steady rhythm of rattling knives, forks, plates and mugs. The room was half obscured by columns of fag smoke and steam, but I could see there was a long queue of men at the counter and wondered if I'd have time to spend my money.

'*Ay-upp, gaffer's 'ere!*' The last man in the queue was looking down at me. '*Wot yuh after, cocker?*'

I looked at the tuffees on offer. '*Milky Way,*' I answered.

The man picked me up and headed towards the counter. *''Utch upp. Gaffer's in a 'urry.*'

The men laughed and allowed me to the front.

'*Wot yuh after, me duck?*' asked the lady in a white nylon tabard.

'*Milky Way.*'

The lady handed me two Milky Ways, but I immediately handed one back along with the coin.

'*Way 'ad tew many delivered,*' she explained. '*Milky Ways don't cost nowt. T'day ownee.*'

'*Ah'll 'ay a cupple o' them free Milky Ways.*' The man who'd carried me over was waving to the lady in the tabard.

'*Yuh can bugger off!*' she laughed. '*Ownee fo' me best custummers.*'

I thanked the lady in the tabard and everyone in the room, but they didn't hear me. There was too much shouting and good-natured laughing.

'*Gaffer's gorriz snap,*' one of them called.

'*Ay's lukkin' pleased wi' 'imsen,*' said another.

And I was. Free Milky Ways and driving a lorry would surely put anyone in high spirits.

Outside, the car park was much busier than when we'd arrived. Two ambulances were stood next to each other and their open doors were surrounded by small groups of men, some in overalls, some in chequered jackets with brown elbow patches. Lying on a trolley in one of the ambulances, I could just make out a bloodied face. There was a lot of movement in the ambulance and the other men were ushered away. They pulled grim faces, took off their helmets and raked angry hands through damp, dusty hair. The leading man threw his helmet at the wall of the canteen and it ricocheted back into the road, rolling over and over until he let fly with his boot. The helmet sailed around the corner and into the grill on the front of Les' lorry.

Les didn't mind. He simply picked up the helmet and held it out while the men walked past. As the owner reached out for it, Les looked at the man's eyes.

The man looked back. *'Ay'll be oaraight burr it split 'is 'ead oppen. Split it oppen like a boiled egg. It's a bad 'un. Ay were bluddy lukky.'*

The lady in the tabard was now standing at the canteen doors. They were open, but the canteen was frighteningly silent. Not even the stray clink of a teaspoon or the scrape of a chair. All the men sat or stood perfectly still. A real-life still-life painting.

In the lorry, driving back to the shed, I asked Les a question. *'Why duh miners laff s' much?'*

He sniffed and thought, then said, *'They 'ay t' laff.'* He sniffed and thought for a few more seconds. *'O' they'd . . .'*

I knew what he meant.

Back at the house, Les walked me to the door again and shouted through to the kitchen. *'Ay's bin wi' may, Hilda. They've tukk Dickie Boom upp t' t' ospital. Ay's oaraight, burr ay's split 'is 'ead oppen.'*

Dickie Boom was a lovely man. Dickie, his son, was my best friend for most of the Pit House years. Sometimes, we were allowed to sleep in an orange plastic teepee in Dickie's back garden. His dad would sit with us, telling jokes and talking about football until we fell asleep.

For the next few days, I didn't see Dickie Jr. He wasn't at

school, which was a relief. I was scared of seeing him and decided to wait until I knew his dad was better. In the end, he came to see me. His face appeared at the front room window and I pointed to the back door. He smiled when he told me his dad was better and that he'd been brought back home.

'*Weer y' gooin'?*' I wanted to know.

'*No-weer.*'

'*Ah'll cumm wi' yuh.*'

Instinctively, we walked towards the Rezzer. It was tadpole season.

Dad might not have said much about mining or our family's connection to coal, but the few bits and pieces he did tell me – plus anything picked up from the Men in the Shed – were enough for me to work out a story of sorts. Our story. I began to understand that serious accidents – like the one suffered by Dickie Boom – happened to miners on a mundanely regular basis. Lost fingers, suffocation, crushed legs that would require the use of a wheelchair for the rest of your life. In my dad's case, mining had left him with partial deafness, countless scars and vibration white finger (VWF: little feeling in the fingers), the latter caused by the vibration of the heavy tools he used. Plus, the recurring dermatitis, which became so bad that he was hospitalised a couple of times.

Mining was also starting to destroy Dad's lungs. Yes, he was a smoker, but not a professional smoker like my mam. Even on his days off, he would have no more than six or seven of his thin hand-rolled fags. Coal dust was the real killer. Unfortunately for my dad, the human body can neither expel nor break down coal dust, so when coal dust entered his lungs, it stayed there. For ever. The result was miner's lung. For my dad's generation – and all those generations before him – miner's lung generally condemned the sufferer to an early, painful and wheezing death. An embarrassing death. Men who had once carved their way through solid earth couldn't even manage to bring in a shovelful of coal from the shed.

I began to understand a lot of things. Like why so many miners never got to celebrate their seventieth birthday. Within a year or two of retirement, they'd wilt, wither and die. One or two of them haunted the Pit Houses, initially happy to be free from shifts at Pye Hill or the Bull & Butcher Pit. Perhaps a year or two of carefree, cloud-nine days – gardening and laughing, invigorating walks to and from the Dog & Doublet – before they quickly faded from view.

Men like Jacko. Mid-sixties, recently retired. Always beautifully turned out in a spotless jacket, crisp shirt, a dandy's waistcoat, trousers fresh from the ironing board, recently polished shoes and a rakish trilby, tilted to just the right amount of lopsided.

Carefree. Cloud nine.

But gardening and laughing would suddenly feel like hard work and invigorating walks gave way to hours spent leaning on the front gate, talking to friends and neighbours. Riding my RSW 11 past Jacko's house, I watched him try in vain to trim his privet hedge with old-fashioned garden shears. His thin arms barely able to lift them, his chest heaving and his red face glowing like a hot coal. I asked if he wanted any help, but he shook his head.

'Ah'm raight, Danny-lad.'

Sadly, he wasn't *'raight'* and a couple of months later there came a second retirement. The garden and privet hedge were now left to their own devices and all of Jacko's talking was done from a comfy armchair in the front room. Or a bed, brought down because the stairs were a bit steep. It wasn't long before the talking stopped altogether. Silence, except for the coughing and the wheezing. Then not even the coughing and wheezing.

So, why did Jacko do it? Why did Dad do it? Why did Selston do it? Why did *we* do it? Mainly because mining was 'our thing', an East Midlands staple for working-class families like mine. Several times, I asked my dad about 'our thing' and, in particular, if his dad was a miner. Several times, he said nothing. I asked again. Again, he said nothing.

But one day, he said something.

'Ah, me dad were dahn t' pit.'

So, I asked about Dad's grandad.

'Ah.'

And his grandad's dad.

'Ah.'

And his great-grandad's dad.

'Ah'.

All the way to my great-great-great-great-grandad John.

'Ah.'

All of them were miners or 'hewers', along with whole hosts of uncles, cousins, neighbours and friends.

The word 'hewer' had largely fallen out of fashion by the time we were living in the Pit Houses, but Dad and the Men in the Shed would sometimes let it drop into the conversation. As usual with words I didn't understand, I immediately went to look it up in the dictionary in the big classroom at school: 'From the Old English *heawan*, which means to hack, chop or cut with a tool.'

Word and meaning now fixed in my mind, I felt the full weight of those five letters. The darkness and heat; the faint but constant crack-crack of the shifting rock above my head. I would imagine myself 'hewing', forcing my basic hand tools into a hard surface. Still trying to get power and purchase from my aching limbs as I neared the end of a twelve-hour shift. Lying on my side like a grisly foetus curled within a claustrophobic eighteen-inch-high womb that provided the only access to a rich seam of East Midlands coal. And when my twelve-hour shift came to an end, I would go home, collecting two or three young sons who worked as door-keepers or wagon boys.

And we would fall asleep.

And we would start again.

In the big classroom at school, there was also a book called the *Reader's Digest AA Book of the Road*. I took it from the shelf, opened it, found Selston and, with a piece of string, scribed a 20-mile arc from north-west to north-east. That patch of land contained all the names of the towns I'd heard about, the places where my dad's relatives had lived, right the way back to Great-Great-Great-Great-Grandad John, who grew up in the small mining towns just east of Chesterfield. Small mining towns where his son, Thomas, clocked on for his first shift at five years old. The same age that I rode the bus to Selston Church of England Infant School reading *The Topper*.

Sadly, Thomas' life story didn't get much better. He lost one son at birth and two daughters, aged three and six. Mind you, compared with Great-Great-Uncle Enoch, Thomas didn't do too badly. Enoch had seventeen children . . . ten of them were dead by the age of two. It's hard to imagine how anyone copes with the loss of one child, but ten? Has evolution armed us with a safety valve by giving grief and sadness an upper limit? The loss of one child must have badly damaged his heart, the loss of two would have easily broken it and the loss of three smashed what was left into a million bewildered pieces. By four, he would have begun to lose count. By five . . . perhaps he no longer experienced any fresh grief or sadness because those feelings were now part and parcel of who he was. They became old habits, like picking his nose or biting his fingernails. He didn't even realise he was sad.

Imagine dangling a foot into the sea. The foot is wet. Then the leg. Then both legs. They are wet. Then you jump in and disappear under the waves. Everything is wet and, no matter what happens or how ferociously the storm rages, you won't get any wetter.

Is that how grief works?

Could I blame coal for those deaths? For Thomas' and Enoch's grief and sadness? Well, I might be able to blame some of it on coal. The hewer's poor wages meant he lived in poor housing with poor heating, poor hygiene, a poor diet and poor . . . everything else. In those circumstances, there was no guarantee that a child born in Selston, Clipstone, Annesley or any other village that was attached to a colliery would make it into double figures.

Great-Great-Great-Grandad Thomas worked at Staveley Pit near Chesterfield, which was made up of several collieries and foundries, all owned by George Hodgkinson Barrow and then his brother, Richard. Thomas and his family lived on what was known as The Long Row, a collection of terraced 'dwellings' that were built by the Barrow family for their employees. Thomas' twelve- and sixteen-year-old sons were miners, too.

Even in Thomas' day, there was no denying that mining was dangerous, and the sheer volume of fatal and serious accidents in the middle part of the nineteenth century had become a subject of national concern. I knew this was true because Ernie Rake, one of the Men in the Shed, told me. He said that between 1850 and 1900 almost 60,000 people died in this country as a result of mining-related accidents. His eyes welled with tears as he explained these figures . . . three miners dying every day. One house overwhelmed by sadness every eight hours.

Maybe you got lucky and the accident didn't kill you. Between 1866 and 1919, a miner was seriously injured every two hours.

Every day of every week.

Every week of every year.

It was when I told Dad about Ernie Rake's 60,000 lost souls that Dad told me a bit more about Great-Great-Great-Grandad Thomas. About how Thomas became one of those 60,000 in 1854 when a corf, a large wooden wagon used to haul coal and soil from the pit bottom, fell 100 yards down the shaft and crushed him. Although I was the only other person at the kitchen table when Dad told this story, he didn't seem to be telling *me*. Maybe he was telling himself. A gentle reminder.

And while Dad was undoubtedly proud of our family's mining roots, that pride seemed strangely absent from anything he said about Thomas or even his own father. Knowing a little about Dad and miners in general, I suspect that pride would have seemed far too garish an emotion for these moments.

As my dad sat, hands respectfully folded in front of him, at the kitchen table, I saw . . . the shadow of something. A tiny chink of darkness. But the hands quickly reached for his Old Holborn tin and the darkness was banished by the phosphorous glare of England's Glory.

As Dad smoked, I thought about Thomas' sons; both working at the same pit. Were they with him when it happened? How did they break the news to their mother? How would I tell my mam?

Not wanting to risk losing their regular wage, Thomas' sons stayed on at the pit, trudging past the spot where their father had died, day after day after day. And their own sons and grandsons carried on trudging, working seams at pits that drew them closer and closer to Selston. One of them was my grandad, Jack. He was a miner, until he was called up in 1914. He fought at Ypres and the Somme, and won some medals, then went back to being a miner.

My grandad's regiment was the Sherwood Foresters and because a lot of those local recruits had mining experience, they were often involved in underground warfare, sinking mines and shafts that would take them under German trenches on the Western Front. Over 11,000 of Grandad's comrades died in the war. Many of them underground. Mining men from Nottinghamshire and Derbyshire, digging in the dark.

In 1923, as a tribute to the men of that regiment, a tower was built at the top of Crich Hill in Derbyshire, almost 1,000 feet above sea level. So high that the people of both Nottinghamshire and Derbyshire would be able to look and remember. From the top, you can not only see Nottinghamshire and Derbyshire, but also Yorkshire, Lancashire, Staffordshire, Leicestershire and Lincolnshire. Even Lincoln Cathedral, over 50 miles away, is clearly visible. A glorious view.

Maybe even glorious enough to make up for how little the Sherwood Foresters saw from the bottom of a pit shaft.

Maybe?

My dad's journey to the coalface, while no doubt written in the same scattered stars that guided Grandad Jack and the rest of his ancestors, came with a slight twist. After leaving school at fourteen, he got a job at the pipe yard – connected to Pye Hill Pit and owned by the same company – making ceramic gas pipes. His offer from the pit came later, via the wartime Minister of Labour and National Service, Ernest Bevin.

No one would argue that going down the pit was as dangerous as protecting merchant shipping in the Atlantic, storming Monte Cassino or dodging bullets on the Normandy beaches. But it was dangerous. A dangerous, disgraceful, bottom-of-the-pile job. Even in a war – even when it became a reserved occupation in 1943 – not many people wanted to be miners.

So few men signed up for pit work that, from December 1943, Ernest Bevin introduced the civil conscription programme for miners. As well as being called up to the armed forces, men between the ages of eighteen and twenty-five could be called up to work at the pit. These were the Bevin Boys.

Men would rather take their chances on the front line than go down the pit. Yes, there was a fair chance you might get killed or lose both legs in the army, navy or air force, but at least you were fighting for your country. You were given a uniform and you won medals for your bravery. People cheered in the street and bought you drinks in the pub. Parents were proud of you.

The only uniform miners wore was a layer of shame and coal dust. No medals for copping a rockfall or losing an eye while trying to save your mate's life. Don't expect a War Disablement Pension either, because it was classed as an industrial accident. Even though a Bevin Boy might lose his life underground (5,000 miners died during the Second

World War), his family would not be given any financial help.

Despite all of that and despite the fact that the law forced them to work down the pit, no one clapped or cheered for the miners and the Bevin Boys. No one clapped or cheered for my dad. No one bought him drinks in the pub. There were no rattled saucepans or hollers of appreciation. There was no glorious homecoming. Instead, the Bevin Boys were beaten up by their (former) mates. Spat on. They were harassed, arrested and accused of desertion by the police. (As many conscientious objectors had opted for pit work, people naturally assumed that anyone who suddenly appeared at Pye Hill, Brinsley or Bentinck in 1943 was a 'conchie'.) Mining families were insulted and ostracised. Poor Mrs Smith was told that her son died in the push for Berlin, just weeks before the end of the war. Imagine how she felt when she saw Mrs Jones, whose son spent his days lounging about on a Nottinghamshire coalface.

Going down the pit – even though the Bevin Boys had no choice in the matter – was seen as the coward's way out. And nobody likes a coward, especially during a war.

Unsurprisingly, most Bevin Boys left the industry after they were demobbed in 1948. A few, like my dad, stayed on. Sitting at the kitchen table, I asked him why, but he was focussed on the rolling of another fag. Perhaps the answer was too obvious to need saying out loud. He'd made a lot of friends during his years underground and he was worried those friends would think him a coward for going back above ground. Worried those friends would think that he considered himself too good for mining. Off to get a nice, cushy, clean, safe job on the buses or the railways.

Miners. Cowards for going underground and cowards for going overground.

Miners. Death, dermatitis and Dickie Boom's head split open like a boiled egg.

Miners. Worried wives knocking at our door and asking if my mam knew where Mr So-And-So was, simply because he wasn't back at his usual time. And the change of tone in my mam's voice if Dad wasn't back from the night shift at his usual time.

'*Ay'll be doin' a bit o' ovvertime,*' she'd say to me and to herself, but I knew exactly what she was thinking.

Skinheads Don't Bother Me

I must have been six or seven. I was in the passage, standing close to the wall and Mam was letting off steam. I was in danger of getting a good hiding, so I readied myself for the nimble boxer routine . . . ducking and shimmying, etc, etc.

That's when my mam added something new: *"F' yuh dunna be'ave y'sen, y' gooin' back t' t' orphanage.'*

As I wasn't sure what an orphanage was or why I would have been there, Mam gave me some of the backstory. Her and Dad weren't actually my real mam and dad. My real (birth) mam and dad didn't want me, so they chucked me out of their house when I was a baby. Then my (new) mam and dad came along and rescued me. (I was adopted, but I don't remember hearing that word. It was always 'rescued' or 'saved'.) If it hadn't been for them (the new mam and dad, but especially my mam), I'd still be there. In the orphan-age-thingy. And if I didn't start playing the game by my (new) mam's rules, that's where I would end up. Back in the orphanage, with all the other girls and boys who had the misfortune to be brought into this world by such cold-hearted (birth) mams and dads.

It was obviously some kind of 'threat', but in my resourceful and curious mind, this orphanage sounded similar to school. I liked school and the threat of being sent there didn't worry me. School dinners were nicer than my mam's dinners, and the teachers said I was good at drawing and singing hymns.

Despite Mam's dire warnings and this other sinister 'mam

and dad', I never gave it or them a moment's thought. I had a home. I went to school. I had an RSW II. I had a mam and dad. Given the choice, they weren't the mam and dad I would have signed up for – an angry blind woman who set fire to stuff and a man who was hardly ever there – but that's the luck of the draw. Yes, I was intrigued by the idea of adoption and extra parents, but that was mainly because my (new) mam kept altering and adding to the story every few months.

There was the original orphanage tale. Then she told me that my (birth) mam was poorly and had to go into hospital, so she couldn't look after me. For a while, I felt sorry for my (birth) mam. Then everything changed and my (new) mam explained that my (birth) mam was very posh and strict, and lived in a very posh house with strict rules. Obviously, a snotty-nosed scumbag like me didn't fit into her very posh, strict world, so she had given me the heave-ho. Every once in a while, my (new) mam would throw in other children that belonged to my (birth) mam . . . several brothers, sisters, half-brothers, half-sisters, stepbrothers and stepsisters. Good God, how many of us were there? Had we all ended up in the orphanage?

(New) Mam filled in a few particulars about my (birth) dad, too. Something about a massive win on the football pools. Failed marriages and affairs. More half- and step-siblings. He was in the navy or possibly a farmer. He was tall. He was small. He wore a hat. He was bald. He was Scottish, Australian, Russian, Polish, Ukrainian or Mongolian. Mongolian? Really?

But even with these garish details, I didn't feel . . . anxious or upset. I didn't spend days and nights speculating. Did it really matter if I was half Mongolian? Descended from an ancient tribe that emerged from the scattered hill towns north of Ulaanbaatar, travelling across the Great Gobi and the Nalati Grasslands, through Tashkent, Uchkuduk, Kazaly, Karabutak and into the Russian heartland, whence my (real) dad caught an overnight coach to Alfreton Bus Station? What if there was a massive pools win and a posh woman who lived in a massive posh house? And I really had got twenty-seven brothers, half-brothers, stepbrothers, sisters, half-sisters and stepsisters? Did any of it alter my tally of gold stars and best-speller prizes? Did it stop Sadistic-Bus-Stop twisting my skin and laughing? Did it bother the lapwings or the Men in the Shed? Did it puncture the tyres of my RSW 11? Did it dull the tools that chopped the coal seams at Pye Hill? Did it prevent outbreaks of dermatitis or stop my mam setting fire to herself?

No. None of this made any difference to who I was or who I thought I was. It didn't matter if I was a miner's son from the East Midlands or a half-Mongolian farmer. It didn't even matter if I had four parents or no parents. I was neither happy nor sad. Neither bitter nor grateful. For all I knew, it happened to every child of a similar age who went to my school. Maybe we'd all gather by the hymn-book stand the following morning and say, *'Yuh'll nivvver guess wot me mam towd me last naight.'* Then we'd shrug, wipe our noses on our long-suffering sleeves, turn to page forty-eight and rattle through 'The Lord's My Shepherd'.

Around the same time that I found out about the adoption, The Tams had a number one hit with a song called 'Hey Girl Don't Bother Me'. As skinheads were also in the news – their arrival at the Rovers in *Coronation Street* in 1971 caused something of an uproar – the two had been combined and the Tams' lyrics changed. People would sing: 'Skinheads Don't Bother Me'.

So, I changed the lyrics as well: 'Adoption Don't Bother Me'.

And it didn't.

For much of the late 1960s and early '70s, my dad and the rest of the miners were at war with the British government. The National Union of Mineworkers (NUM) was not the biggest union in the UK, but it was unquestionably the strongest. So, in 1971, the NUM flexed its considerable muscle by asking the government for a 43 per cent pay rise. Although the public sided with the miners, the government turned them down and, in January 1972, the miners went on strike – the first nationwide official strike since 1926. This was headline news. Britain ground to a halt, the government declared a state of emergency and there was even talk of a military takeover in Downing Street.

Against the combined might of my dad, Dickie Boom's dad, Les and their mates, the country would fall.

To conserve coal reserves, factories and schools shut their doors and power cuts plunged Selston, along with most of the country, into darkness. Which, of course, gave our household a distinct advantage. We were used to the darkness. The fact that we now had a valid reason to live in this can't-see-your-hand-in-front-of-your-face pitch-black world turned our house and its darkness into a kind of game. It was entertainment. There were candlelit dinners and paraffin lamps. Torches and long, scary shadows climbing every wall. I found an old Parker Chalwyn Roadworks oil lamp that filled my bedroom with an eerie orange glow. Groups of children gathered at the top of Jubilee Hill and looked west, over towards Somercotes, Alfreton, Riddings, Swanwick and Ripley, marvelling at how dark the world was and how little we could see. Waving at passing motorists as they caught us in the glare of their headlights.

Even the ridiculously cold weather – not far off minus 20 some nights – didn't dull my enjoyment. As Dad was a miner and we lived next door to a pit, the coal shed was always well stocked, allowing Dad to keep both fires going for most of the day and night. Sensing that a strike and power cuts were in the air, he'd also bought one of those brew-at-home beer kits and filled several large buckets with his own foaming, earthy syrup.

As far as I was concerned, the biggest difference the strike made to those days was that Dad became part of them. It took a week or so, but I eventually got used to him being there. At home, with me. Not going to work at night. Not opening and closing the kitchen door just after nine o'clock. Not striking a match in the garden or cupping his hands to light one of his hand-rolled fags. Not strolling down the garden, opening the gate and heading towards Pye Hill. I was no longer watching the gate close or watching the tiny orange glow of Dad's fag. Dad's fag wasn't joined by the tiny orange glow of more fags, bobbing and weaving like insects in the wind. No sign of at least three or four other men from the Pit Houses who were also on nights. Dad and the other men weren't walking together, smoking. No sound of their heavy, unhurried boots or the brief, murmured greetings.

'*Ow do.*'

'*Ay-upp.*'

'*A' tha raight?*'

'*Non bad.*'

'*Middlin'.*'

Even the coughs had disappeared. *COOOUUGGHHH!* (That bloke from Pye Bridge.) *Hmm . . . hmm . . . coff-coff.* (My mate Graham's dad.) And my dad's coughs. Single, distinct coughs, one after the other. As if he was counting. *I have seven apples and I take away three. How many apples do I have left?*

Cough. Cough. Cough. Cough.

These were the sounds and images that had lulled me to sleep for ever. And now, thanks to the strike, they were gone.

Although Dad was no militant miner, NUM solidarity was strong and, most importantly, the strike was 'official' – a nationwide ballot had voted 58.5 per cent for strike action, just above the 55 per cent threshold. It meant money was tight and my mam had to cut down on her fags, but breaking the strike or questioning its legitimacy in a small mining community like Selston was not an option. The men had voted. Everybody out!

This was the first chance I'd had to witness Dad as 'Dad'. What time did he have his breakfast? When did he have a bath? How did he keep himself busy? Like a botanist observing a rare orchid in the rainforest, I took notes: went for a walk, did some bits in the garden, *'ad' a nod* (sleep). Stoked both the kitchen and living room fires, talked to the other miners that lived in the other Pit Houses and had his tea about half past five.

Plus, the grand finale. A pint of his foaming, earthy syrup – just a single pint, mind, nothing more – poured into a dull silver tankard. Having stuffed a poker into the front room fire in preparation, he would then pull up a scorched armchair, plunge the poker into the beer (popular on cold nights when pubs with real fires would provide a selection of pokers) and sit contentedly, sipping and reading his book by the combined light of fire, candle and paraffin lamp.

Apart from a newspaper, I'd never seen my dad read anything before. I sat in the other scorched armchair, hutched it towards him and asked what book he was reading.

'*Nowt,*' he answered, turning to the side.

He was obviously reading '*summat*' and I could just make out the red letters on the black book cover: Dennis Wheatley. And above, in white: 'A Black Magic Story'. And below, in white: *The Devil Rides Out*. And below that: a picture of candles and fire.

Candles? Fire? Black? Obviously, something to do with coal and the strike. Was Dennis Wheatley the president of the NUM?

This became a regular appointment. After we'd finished tea, Dad would settle himself with his beer and I would settle myself with a glass of poker-warmed Dandelion & Burdock – thankfully, the Corona Pop Man was still doing his rounds. Dad would read his Dennis Wheatley mining history books and I would read the Bible. Not reading it with any of that hell-and-damnation-and-resurrection-and-Holy-Trinity palaver. I just liked the words and the pattern of those words, tying them to a rhythm, like the lyrics of a song in the Top 40.

The more I read, the more I sensed a bizarre familiarity with some of the Old Testament bits. What with all the 'hath', 'hast', 'thee', 'thou', 'smote', 'brethren', 'unclean' and 'do not mourn', it reminded me of Dad and his mates or the Men in the Shed. Soon, I was translating whole verses into Selston-ese, chuckling silently to myself as I shifted the location from the Middle East to the Middle East Midlands.

'Then, Dickie Boom gorrupp early in t' mornin', an' set aht on t' main road, crossin' ovver t' Erewash an' int' t' land that summ call Timnath-serah and summ know as Somercotes.'

Mam often went to bed early during the strike and after she'd finished shuffling about upstairs – soaking her dentures, a tiddle in the po – a warm, comforting calm settled on the house. Leaning back in my scorched armchair, I would take a sip of Dandelion & Burdock and listen to the fire, watching the reds and yellows flash and waver across Dad and his book, casting yet more of those agitated, tricksy shadows on the walls and ceiling. Occasional cars, motorbikes and lorries went past the window, but the gentle air in the front room cushioned and muffled the sound. By the time I heard them, the engine noises were on their way to somewhere else . . . passing the crossroads at the Top 'n' Town, idling past the Dog & Doublet.

Despite my age, my poor haircut and uncertain adopted history, these evenings left me feeling worldly and content. I was proud of what I'd accomplished. Knowledge had become my armour, with facts, figures and words hanging from my body like chain mail, warding off evil spirits and shielding me from life's crushing blows. Making me bulletproof.

But sitting there in my scorched armchair, legs too short to reach the ground, Dandelion & Burdock almost finished, I also savoured the joy of time spent away from facts, figures and words. The gift of silence. The benefits of nothing. I had a vague feeling that, in that silent nothing, I was tuning in to my dad's wavelength. Like a crow or a magpie, settling on one of the thousands of electric and telephone wires that intersected the lower levels of the Selston sky. The wires themselves made no sound, but they carried power and conversation, assets and information. They carried life. While the front room and scorched armchairs made no sound, they, too, were full to bursting with images, emotions, ideas and understanding. Hundreds of thousands of massless particles bouncing between me and Dad. A dialogue . . . of sorts. Our dialogue.

I added it to the list that Dad and me shared – Rezzer, lapwings, blackberries, mending the fence, D500 drill, cow parsley blowpipe, Dennis Wheatley, the Bible, Dandelion & Burdock, beer, an unspoken dialogue – and hoped that the strike and the darkness and the peace would last for ever.

Sadly, it was all over by the end of February. The miners hadn't got the 43 per cent they wanted, but their new wage packet was one of the highest of any of the UK's working-class industries. I missed seeing Dad doing normal stuff, but he seemed happy to be going back to work. Happier about that than he did about the extra money. Mam made up for this by being very happy about the extra money and sent me down to the shop in Pye Bridge to get twenty Player's No. 6. Her celebrations continued for several days and she decided to spend one evening finishing the bucket of Dad's earthy, foaming syrup that sat in the corner of the kitchen. After half a dozen glasses, she started banging on the wall, loudly beckoning Vera from next door. Mam was now crying and kept repeating, 'Ah luvv 'im. Ah do.'

As Vera positioned one of the empty buckets for my mam to be sick, I decided to go to bed, but it wasn't long before Mam made her way upstairs and bundled into my bedroom. Feeling for the edge of my bed, she sat down.

'Ah've seen t' Devil, y' know.'

So, that's who she was in love with . . . the Devil. Was it the one who 'rode out'? The one from Dad's Dennis Wheatley book?

'Ah've bluddy seen 'im!' She slammed her hand onto the bed. 'Blue Devil. Ay's all blue. Ay's gorra blue suit, like y' dad's. An' black hair like Elvis. An' ay's tall . . . ever s' thin.'

And with that, she stumbled into her own bedroom.

She was still asleep when I woke up, so I made some

Weetabix and went to school, keeping my eyes peeled for a tall man in a blue suit. Was Mam talking about *the* Devil? The one from the Bible? Must be. He was obviously here, in Selston. How else would a blind woman have been able to see him? This Devil had granted her the gift of sight for a few minutes because she said that she loved him. And if she kept on loving him, she'd be able to see again.

It took a few days, but I eventually spotted him. And once I did spot him, I couldn't stop spotting him. There he was, by the Co-op, getting into his car, a white Vauxhall Viva. And playing darts at the Bull & Butcher, winning every game. Cleaning the windows outside the council offices at the bottom of Buxtons Hill, making a lovely job of it. He was also spotted by wise and honest people like Mr Greaterex, Mrs Bullock and Ow'd Man Wilson, who used to run the post office. Blue suit, black hair, tall and thin. Oh, yes, it was him.

I needed to get close. I had . . . questions. So, each night, I flew. I flew and flew and flew, ranging past points I'd never passed before, sinking down to street level, searching houses and back yards and the churchyard, looking for him and for answers. Why my mam? What had she done to catch his attention? Or was it Dad that the Devil was interested in? Did it all have something to do with that book, *The Devil Rides Out*? Had Pye Hill's deep shafts cracked open the Gates of Hell? Was he after me? Did I really earn all those Clever Bugger stars and prizes by myself or did I have help? Had the Devil stepped in like a kindly uncle, offering encouragement and sanctuary to a deserving young lad?

I almost had an answer later that month. Down past the Clay Heaps, where the clay finished and the grass started, there was a stream. After recent heavy rains, the stream was at its best, a crisp, clear fountain springing from nowhere. Out of a hole in the heathery grass, then away on a wiggly plunge past the pit and further downfield, cutting slow, dark pools deep into the earth, then squeezing upwards into glassy, much speedier shallows. Some of the slow, dark pools were as big and deep as a tin bath. Skirted by arrogant clumps of bacon and eggs (bird's-foot trefoil), daisies, dandelions and buttercups that would dazzle my eyes at noon.

As I walked, I searched for rocks, branches and bits of brick, damming and diverting the water, parting the miniature waves, scraping fresh channels to take the non-stop ripples past this gorse bush and that thistle. As if dragged by an unseen force, I began to follow the stream back up to Barrows Hill Lane, where I climbed over a fence and onto the road. That's when I saw him. Underneath the bonnet of his white Vauxhall Viva, which itself was parked under a large ash tree. Still in his blue suit, poking at bits of the engine.

Having picked up various hints and tips from the 12th edition of *The Motor Repair Manual*, I felt confident enough to stride over and tell him that most Viva problems were caused by poor cleaning and lubrication of the ignition timing mechanism. Sure enough, he pulled off the distributor cap, fiddled about a bit, put back the distributor cap, tried the engine and . . . started first time.

'Yuh owe may one,' I laughed.

The Devil smiled and winked. '*Cheeky bugger. Wot a' tha after?*' he asked. '*All t' kingdumms o' t' wold an' their splendour?*'

I shook my head. '*Nowt like that. Ah just need t' know abaht me mam and me dad. An' t' miners.*'

'*Wot d' yuh want t' know?*'

'*Ah'm non sure. Ah need t' think abaht it,*' I said. "*Ow long 'ave ah got?*"

'*Tekk y' time. Ah'll sithee when tha's reddy.*' We shook hands warmly as he folded himself into the Viva and sped away towards Jacksdale. Understandably, he didn't want to miss the big lunchtime darts match at the Welfare.

Rediffusion

Somebody was dead and my mam seemed pleased. The fact that this person – a man, vaguely related to us – had died meant we could move into the council house he'd just died in.

It was as simple as that. We left the Pit Houses sometime in the spring of 1973. Inside, I'd always thought of it as a 'full' house with lots of stuff, but when it came to the actual day of the move, our lives fitted easily into the back of a van, with the bedsteads tied on top. The reason we didn't have much to take with us was because we left a lot of it at the Pit House. I heard my mam saying something to Vera about my dad getting a pay rise – the 1972 strike that led to the blackouts had ended after the miners were given their almost 30 per cent pay rise – and we were going to buy some new furniture from a shop in Somercotes.

There were a few people I wanted to say goodbye to. Dickie Boom, although we would still see each other every day at school. Duggy Dugg, although we would see still each other every day at school. I had grown fond of Duggy Dugg, who lived in a small terrace on the other side of the road. He didn't have a dad, but his mam laughed a lot and opened her eyes wide when she talked. She used to let us play records on her record player. Lindisfarne, Rod Stewart and *20 Dynamic Hits* by Various Artists.

And I needed to say a fond farewell to Sadistic-Bus-Stop. Although I no longer saw him every day at school – he'd moved up to primary school the previous September – we were still neighbours and he continued to torture me outside office hours. Our new house would put a stop to that. As a celebration, I tore out several pages from my Bible and highlighted words that made up sinister sentences – 'God has spoken. You shall die', 'Vengeance and Pestilence. It is written!' and so on – and posted them through his letter box.

With a light and airy spring in my step, I then said goodbye to the people and things that really did matter. The Rezzer, sitting on the fence, looking out over its evil water. I was being too harsh. Not evil . . . just unforgiving. Like my mam, the Rezzer took no prisoners. The grass was shorter now, but I could still picture Football Shirt's snow-white goosebump skin. I could still picture frogspawn, tadpoles and frogs. Older boys throwing those frogs over towards the main road or slamming them against tree trunks. I tried hard not to remember that.

I said goodbye to the honest lapwings. A handful of adults and this year's youngsters were picking and pecking their way across a length of scrubby grass at the bottom of the field. My shouting lifted them from the ground, but they could see it was only me and were soon back in position. About halfway down the field, I knelt and felt the earth with both hands, picking up clumps and crumbling them with care. Making sure the worms and beetles were unharmed. I thanked them all. Not just for the special days when Dad and me were walking to the pit to get his wages, but for the countless mornings, afternoons and evenings when I stood on the back yard. And watched those lapwings and crows and sparrows and greenfinches and blue tits. And smelled the soil. And worshipped the rosebay willowherb. And wondered how far away the hills of the Peak District really were.

Walking down the field, I eventually reached the edge of the pit ground, gently stroking grasses and hedges and the stout cow parsley. It was all this that I'd miss most of all. It had been the backdrop to my world, the picture on every wall. For all of my short life, Pye Hill Pit had held me securely in its arms, providing me with moments, hours and days to treasure. As long as the lights shone, the smoke billowed and the men laughed, then everything was safe. Me and my future. The Pit Houses and the whole of Selston. Dickie Boom and Duggy Dugg and everyone I had ever set eyes upon.

To know Pye Hill was to know that our world would plough on. My great-great-great-great-grandad John had worked these very same coal seams. His relatives, too. And even if his relatives weren't really my relatives, it changed nothing. I had been grown, fed and watered in this earth. Look, there were my footprints and my memories, as much a part of the landscape as any plant or animal or cloud or lump of coal.

I said goodbye to Jubilee Hill, always ready to guide me towards Pye Bridge and the corner shop. Goodbye to the River Erewash and the short tunnel that led to a label on a pot of jam.

Dad was still working at 'the pit', but now he would be working at a different pit – formerly known as Selston Pit but recently renamed Pye Hill No. 1; the old Pye Hill near the Pit Houses was now known as Pye Hill No. 2. That also meant saying goodbye to the Clay Heaps and the Men in the Shed. Goodbye to their wise words and warm laughter. One of them gave me a present, a large hunting knife carved out of blonde wood. Too blunt to hunt but perfectly shaped to dangle from a snake belt.

Now I was ready for the journey.

For the first mile and a bit, the van and the bedsteads followed the same route as the bus that took me to Selston Church of England Infant School. Up Alfreton Road to Top 'n' Town, sweeping down past the first Co-op and where the old Bull & Butcher Pit used to be, past the Bull & Butcher pub, the council offices, up Buxtons Hill, past the chapel, the second Co-op and the Crown pub, past Toad Hole and my school, past another chapel, looping down again past the Bottom Rec and Allen's Green to Wilde's Corner and the petrol station. As far as I was concerned, this was all . . . new. And some of it seemed much too nice to be Selston. Large detached houses and bungalows, layered gardens, trees with rope swings and cars in every drive. How big had my dad's pay rise been? Were we millionaires?

Staying on that main road took you past a small police station, another garage, a chip shop, the Manor pub and, up the road next to the Manor, a third Co-op. But we turned left at Wilde's Corner, up Mansfield Road for a bit – more nice detached houses and bungalows – then past a scrubby, gruff-looking bit of waste ground. That was more like Selston. Turning right. Definitely more like Selston.

The White City Estate, so called because the grey, sooty houses had indeed been white when the estate was built in 1950. As estates go, it wasn't bad. Half a dozen streets of identical grey/once-white prefab concrete boxes that became two semi-detached homes for people like us. Although these weren't technically Pit Houses, most families on the estate were still connected to the pit in some way.

Inside, each semi-detached pair of houses had the same layout, mirrored from left to right. The front door opened onto a passage that led straight ahead to the kitchen. Off to one side of the front door was the 'front' room; next to the kitchen at the back was the 'back' room. On the other side of the front door, a set of thirteen stairs carried you to a landing and four doors. Bathroom, 'back' bedroom, 'front' bedroom and a smaller box room. From the kitchen, you could also access the back yard, where you'd find an outside lavvy and the coal shed. All miners were given a certain amount of free coal as part of their wages. It arrived every two or three months, stacked in heavy hessian sacks on the back of a lorry. The coalman driving the lorry always wore a cap, heavy-duty gloves and a thick leather jerkin, but his back was also protected by an even thicker layer of leather that seemed to hang from his shoulder. A layer of armour that sat between his spine and the knobbly lumps of coal.

If I wasn't at school, his arrival was always worth a watch. A flatbed Leyland lorry with hinged side panels, stacked with sacks, their rope handles poking up like ears. On the estate, the lorry could park right outside the house and I would immediately tear up the steps, ready to help. As well as picking up any coal that was dropped on the way to the coal shed, I was allowed to carry the empty sacks back up to the lorry, flinging them up to the coalman's mate who stacked them just behind the cab. Ours wouldn't be the only delivery on the road. A lot of miners meant a lot of coal, but as soon as the lorry moved to a different house, I kept my distance. You didn't pick up coal that was meant for another house . . . that was the law.

The 'front' rooms of the estate houses faced the road, which automatically meant they had the best curtains, best ornaments and best pictures on the wall. If there was a telly (we acquired a coin-operated Rediffusion black-and-white set for the new house but, even in 1973, not every house had a telly and many, like ours, were still black-and-white) it was in the front room. On display for everyone to see. I didn't care *what* I was watching; if it was on, I watched it. And that was all that mattered . . . people could walk past our house and see that we had a television and it was switched on.

'*Lukk, way watching* Songs o' Praise . . . *frumm Fisherwick Presbyterian Church in Belfast.*'

The other reason we moved to the estate was because the new house – the one where the man had died – was three doors down from Aunt Lal. She was another of those aunties who wasn't really an auntie and Mam proudly announced that Lal was her best friend. To my knowledge, I had never met Aunt Lal and couldn't imagine my mam having a best friend. But there she was, living three doors up with two daughters, both older than me. According to my mam, Lal's husband had '*run off wi' annuther wumman*'.

It was summer and as soon as school was finished for the holidays, I slipped the hunting knife into my snake belt and set off to explore this new world. For several days, I did nothing more than wander up and down the estate, getting my bearings and staring into windows. These houses were bigger than the Pit Houses and they had large Crittall windows at the front that allowed me to see inside. So I stood at the front of various houses and I looked inside. Sometimes, people would open the front door and tell me to 'Bugger off', some would draw the curtains, others would smile and wave. Then tell me to 'Bugger off'.

There were more children on the estate than in the Pit Houses and there seemed to be a lot more girls. They would gather on the section of grass that marked one end of each street, skipping and shouting and laughing. Meanwhile, the estate boys would gather on the section of grass that marked the other end of each street, playing football and shouting and shouting some more.

As a newcomer, I was not invited to play football, but I hung around anyway, hoping one of the boys would ask me to join in. They didn't, so I walked to the other end of the street and hung around watching the girls skip. That was even worse. They pointed at me and my shorts and called me 'Knobbly Knees'. One of them asked me where I was from. Trying to be clever, I said that I'd been born in the same earth that they were now frolicking upon, which only resulted in more pointing, more 'Knobbly Knees' and much more laughter.

I went home and waited.

And waited.

Six or seven days in total.

Enough time for them to forget.

But I didn't.

And then, one evening, I decided to dress for danger. Was it time for The Fox or The Shadow to come out of retirement? What good would they do? This needed a man of action . . . a new hero? For an hour or two, I toyed with Wonder Boy. Dad had given me a length of thick fishing line and some lead fishing weights, and I experimented with a sort of bolas. The kind of weapon that Wonder Boy might carry. It was rubbish.

What about The Mystery Man? No superpowers, no super-weapons . . . just a man, inhabiting the primeval night, solving crime and righting wrong with his colossal brain. I definitely needed a hat and found a battered flat cap in our new coal shed, probably belonging to the man who died. There was also a longish overcoat and a scarf to keep the hat company. I was a big eight-year-old and the dead man must have been a small dead man because the coat almost fitted me. Trousers were from Dad's wardrobe, turned up to clear my plimsolls. Scarf around the neck? No, worn like an outlaw, covering the lower half of my face.

I crept across the back gardens of all the houses at the top of our street, then left down the hill, still inside the hedge. It was early evening, but the sun was low and the shadows were long enough to hide anything that needed to be hidden. Lying in someone's garden, I could see under the thick hedge and spied a group of young girls playing games and skipping in time with the passing clouds.

I'd give them 'Knobbly Knees'.

And with that, I leapt up and rolled over the hedge.

'Yaaaagghhh!'

Now I had their attention. Stepping into the circle, I used a husky voice to say something, but even I couldn't hear it above their screams.

So I pulled out my wooden hunting knife – coloured black with felt-tip pen – hoping to quieten the ruckus. Opposite effect, really. All of them howling wilder than before, skittling away to tell mams and dads.

'It's made of wood,' I shouted, but nobody was listening.

Luckily, I was back home and in bed, with the knife safely stashed in the coal shed, before the trouble started. I think I got away with it. And even better . . . the boys invited me to play football. Probably worried that if they didn't invite me to play football, I would stab one of them.

Being so close to Lal brought about real change in my mam's life. Her focus began to shift away from me, Dad and her unremitting fury, and on to other things. Such as going to the shops. I'd never seen my mam go to the shops before. Back in the Pit Houses, the only time she went outside was to hang washing in the back yard or clear snow from the side of the house. I'm assuming she did step over the garden boundary at some point, but I don't remember it. She certainly never went down to the shop in Pye Bridge or to the market in Alfreton.

So, the first time she announced that she was going to the shops, I naturally thought she was making it up. She was blind, how did she even know where the shops were or what they were selling? Because Lal was with her, that's how. Lal would walk to the shops, arm in arm with my mam, telling her what was on the shelves: tins of beans, washing powder, 5-amp fuses and the like. I was so surprised by the shop visits that I would often follow them down there, just to see if that was where they were really going. It was strange to see Mam looking so . . . ordinary. Reaching out for a packet of tea or patting a cauliflower.

And there, in the background, me shaking my head with disbelief. A cauliflower!

After 'going shopping', Mam and Lal would walk back, occasionally stopping to light fags or talk to someone. Or talk to someone while they were lighting fags. (Lal always lit Mam's fag for her; she'd obviously been warned about the potential fire risk.) If I stood close enough, I could hear my mam having regular conversations. With people. About the price of lard or some scuffle outside the chip shop.

After watching this for a couple of weeks, I became aware of a new idea forming inside my mind. My mam was no longer just 'my mam'. No longer the 'mam' I had known thus far. She now had connections with the outside world; she wore tights to hide her still angry, scabby, swollen legs; she carried a handbag. It was a bit like seeing one of your teachers in a situation that didn't involve them being in school. They had a real life and did the same sort of stuff that people in Selston did.

And now my mam was at it. Doing the same sort of stuff that people in Selston did.

Something else happened in the new house, too. We had visitors. I'm not talking about those mad, thieving fuckers who used to turn up at the Pit House. These were normal people. Relatives of Aunt Lal's. Other neighbours. A lovely bloke from the furniture shop in Somercotes who came to collect weekly payments for Mam's new dining table and four chairs.

'Lukk at that,' she'd purr, stroking the top of the table when anyone or no one was listening. 'Luvvly, in't it. Tip-top, that is. Tip-top.'

My favourite visitor was a man called Roger, some sort of combined door-to-door salesman/fortune teller/man about town. Roger smoked almost as much as my mam and always arrived with a couple of large, battered suitcases. One week, it might be full of men's singlets and socks or ladies' perfume. The next, it would be packed with toys that looked a bit like the ones that were in the shops or plastic plates and mugs that you could take on a camping holiday.

Roger was a small man with a slow, lazy voice. 'Thisss perfume's frommm Fraaance. It'sss all the rrrage in Parrrisss. Oooh-laa-laa!' Occasionally, if he had an important announcement, he would finish off with a lyrical flourish and a nifty wave of his fag. 'I can dooo you fourrr pairs of worrrk socks . . . FORRR A POUND!' A tiny river of smoke would squirt up to the ceiling, courtesy of the wafted fag. His smile seemed genuine, full of teeth and bright eyes, like the men in tuxedos who danced on Saturday evening telly shows.

I liked Roger. I could have lived without the fags and the plastic plates but was intrigued by the fortune-telling stuff. Like me, it seemed that Roger had found a gateway to another fantastic world. So, when Roger offered his smile and his teeth and his eyes, I offered mine back. And I focussed my thoughts . . . 'Roger, can you hear me? Are you receiving?'

If he heard me, he didn't let on. Probably for the best.

Roger's sessions were always midweek and included Mam,

Aunt Lal plus three or four other women who all crowded around the coffee table in the front room. Mam would let me sit with them while he read palms and tea leaves and tarot cards.

After a couple of hours, the room would be thick with fag smoke, which only added to the ominous tone of his tea-leaf revelations.

'Now, Jean, caaan you tell meee if you've haaad some baaad news, recently?' Roger wanted to know.

Jean would concentrate. Really, really concentrate. Then she'd nod her head. 'A bloke 'oo wokks wi' me 'usband 'as just 'ad 'is leg off.'

'That'll be it, Roger!' Another woman started to nod her head excitedly.

'Well, ay anna 'ad it off yit, burr ay's gooin' t' see t' doctor next wikk,' explained Jean.

'To haaave hiiis leg offf?' asked Roger.

'Ah dunna know,' answered Jean.

'That'll be it, Roger.' The other woman kept nodding.

My mam and Roger had some sort of 'thing'. When I was searching through her dressing table a couple of years later, I found a letter from him. The only reason I looked at it was because I couldn't understand why my mam would keep a letter when she couldn't read. It said something about how he and his mate had enjoyed meeting up with Mam and Aunt Lal.

There was also another bloke who lived down the bottom of the estate. His name was Mr Walters. Occasionally, Mam asked me to walk her down to his house after Dad had gone to work. Then I'd be sent home and told to come and fetch her in two or three hours, by which point she was falling-down drunk.

One time, instead of taking her straight back to our house, I cut across a path that led off the estate and into a field that was home to a couple of horses. Mam had no idea where we were. My plan was to stand her next to the horses and then poke them with a long stick until one of them kicked her in the head. But every time I got within a few feet of them, they trotted off into another part of the field.

Eventually, Mam realised something wasn't right and started asking me where we were. I told her we were on the lawn in the front garden of our house. She wavered a bit, sat down and went to sleep. One side of the field bordered the main road and the orange glow of a modern lamp post turned her slumped figure and everything else an unnatural greenish-orangey-brown. Hardly any of the field was flat. It ran from the crossroads at the White Lion all the way down to the edge of the estate in a series of badly thought-out fits and starts. Nettles, thistles, gorse, grass, dandelions, dock and not much else. It was far too uneven to mow, but the horses – and occasional cows – kept the grass in check.

Damp in the late-night air, the grass and ground smelled heady and much too sweet. As if it had been fertilised with Tooty Frooties and bottles of pineapple Cresta. I sat down next to my mam and stayed with her until she woke up, then took her home and put her to bed. Luckily, she didn't remember the field, the horses or falling asleep.

Or the pineapple Cresta.

Newfangled Ideas and Fresh Thinking

Getting a telly was a big deal. Like my mam with her dining table and four chairs, I felt like standing next to it whenever somebody came to the house, giving the hefty box a reassuring pat and listing all the programmes I'd watched.

'*Did yuh see that doc-you-men-tree on t' EEC?*' I asked Mr Parkin the newsagent when he caught my eye one Friday while collecting his weekly money. He nodded but I could tell he wasn't interested.

My mam heard me talking to him and was keen to intervene. '*Ay watches summ bluddy rubbish,*' she explained. '*It's all talkin'!*'

The telly had indeed blown open the '*It's all talkin'!*' floodgates. Programmes about the design of car engines or the history of lacemaking in Nottingham. The Hindi and Urdu delights of *Nai Zindagi Naya Jeevan* on Sunday morning. Sunday afternoons at twenty-five past five . . . Bronowski's *The Ascent of Man*. A telly was like having school lessons all day, every day.

My mam was hypnotised by the telly. Problem being that she couldn't see it. So, I was called in as an interpreter.

'*Wot's 'appenin'? Wot's ay doin'? 'Oo sedd that?*'

Easy to answer if we were watching Eamonn Andrews introduce the friends and family of Christopher Lee on *This Is Your Life* but not so simple if she insisted on sitting through Alan Plater's *Land of Green Ginger* or Dennis Potter's *Only Make Believe*.

The coin-operated black-and-white telly was not alone.

Next to it sat a Fidelity record player, ordered from the clubby at the cost of £1.16 a week. Mam, Dad and me even made a rare family trip to Woolworths in Alfreton to start our record collection: *On Stage*, a live album by Elvis for my mam; Waylon Jennings' *Ladies Love Outlaws* for Dad; and *20 Power Hits* by Various Artists for me.

Although records were expensive, luck was on my side in 1973. Our new house was also just a ten-minute walk from Selston's brand-new library and, alongside thousands of books, there were several racks of records that could be borrowed and played to death. The soothing, shifting words and guitars of Cat Stevens' *Catch Bull at Four* – brought home because I wanted to know why the bull had to be caught at four. The soundtrack album to the ATV series, *The Strauss Family* – turned up loud, 'The Blue Danube' could rival the majesty of Black Sabbath's 'Spiral Architect'. David Cassidy's album, *Rock Me Baby* – I didn't cry often, but always did when I heard 'How Can I Be Sure'.

I soon became a familiar face at the library, greeting the staff with casual hellos. Sharing a joke about how long I'd borrowed my favourite book, *Know The Game Table Tennis, Published In Collaboration With The English Table Tennis Association*, or drawing their attention to the strange picture on the cover of *Led Zeppelin IV*.

The library was warm and airy. It was quiet. Comfortable, soft chairs were dotted around, begging to be sat on. So I sat. For hours and hours. Usually reading books, but not always. Sometimes, I read the covers and inner sleeves of records. Sometimes, I just sat, not reading anything. Looking. And smelling. Yes, most of all, I smelled.

For the first time in my life, I was in a building where people weren't allowed to smoke. Not even Selston Church of England Infant School had been completely smoke-free . . . some of the teachers smoked, not to mention the eight-year-olds who used to nip into the lavvy for a crafty drag. Stepping into the library was a fresh, full-on assault of the nostrils – the newness of the building and furnishings, books, records, Windolene, the long orange curtains, carpet tiles, more books, more records, instant coffee, pens and pencils, even more books, even more records, a particular perfume worn by one of the well-heeled assistants blending with the chemical cologne of the photocopier. And books.

I didn't even need to steal these books. All I had to do was smile as I produced my library card at the counter and *Heute Abend*, a German language textbook by Magda Kelber, was mine for three weeks. Alongside all the usual stuff about going to the shops and asking for directions, it had a swearing section. How to swear and curse in German.

'*Get dahn t' Wildes an' see if they'll let yuh hay a cupple o' fags,*' shouted my mam. '*Tell Stan ah'll pay forr 'em next wikk.*'

'*Mutter, warum verpisst du dich nicht.*'

Along with the new telly, new record player, new library, new smells, new language skills and new house, there were new people. New people on our new street, like Mr Keith Kent and his rarely seen wife. They lived opposite our house and up a bit. I first noticed Mr Keith Kent because his garden wasn't up to much. Most of the gardens on our street were well kept, neat, tidy and 100 per cent weed-free. Mr Keith Kent's garden, along with his house and hedge, seemed . . . scruffy. Scrabbled together in a hurry. The man himself, Mr Keith Kent, had a loose body and thinning hair, see-through skin that surprised me with its red and blue lines. His shirts were light and floaty, white and stripey, always opened onc button too many. His menswear-department trousers hid urgent legs that stalked up and down the estate, always going and coming back from somewhere. There were broken children's bikes and scooters lined up against the front of Mr Keith Kent's house, but I'd never noticed any children.

And, more to the point, hardly anyone had ever seen his wife.

Mam came to her usual conclusion. '*Ah rekkon shay's run off wi' annuther bloke.*'

With such mystery lurking so close to home, I couldn't

help but take notice of Mr Keith Kent and His Invisible Wife. A glance every time I walked past his scruffy gate and hedge. Sometimes, he'd catch me glancing and I'd dip my head below the hedge and run till I was out of sight. I started watching him from my bedroom window. Watching how he came out of his front door. Looking for clues or signs of a struggle.

Finally, and with a dogged determination to crack this case, I waited for the night and another appearance from . . . The Mystery Man. I slipped out and up our steps, across the road, up a few houses, through his gate, commando-style up the front lawn to his front window. The curtains were drawn but the small slit in the middle was enough for me to see through. A large lady in an armchair, and next to the armchair, a large wheelchair. Mr Keith Kent and his wife were laughing at something on the telly. He stood up, stroked his wife's hair and bent to kiss her on the forehead.

I liked Mr Keith Kent.

I also liked Mr Gregger Kirk. He lived round the corner from Aunt Lal with his wife and teenage son. When Mr Gregger Kirk found out I was a keen birdwatcher, he invited me up to his back garden to see his aviary. I told him I wasn't completely sold on the idea. All those beautiful birds locked in a cage.

Despite the disagreement, time with Mr Gregger Kirk was always time well spent. Now at the same pit as my dad, he'd previously trained as a carpenter and taught me how to hold a saw and a screwdriver. He told me about different wood joints like half-lap, mitered butt and dovetail. And different screws like slot, countersunk and panhead. We made a bird box together and I hung it in the back garden.

And Mr Alf Morton who lived in one of the red-brick houses at the other side of the road at the top of our street. Another miner and another man, like Mr Gregger Kirk, who was good with his hands. So good that he'd made a tank! The body was made out of wood, built around the chassis and engine of a Reliant Regal. For most of the year, it sat under a large tarpaulin in his back garden but came out of hibernation every summer for Selston Carnival.

One Saturday in the summer holidays, the Carnival brought Selston to a standstill. There were marching bands, kazoos, drums, sequined sashes, horses, Boy Scouts, Girl Guides, penny-farthings and drays – flatbed lorries embellished with themed backdrops and costumed characters. Several drays were based on mining, with young lads wearing flat caps, pretending to shovel piles of coal. Some drays anticipated the holiday season, with fruity-looking men done up as women, lying on piles of sand.

I was dragged along with Mam and Aunt Lal to watch the procession wend its way past the bottom of the estate and up towards the large playing fields at Selston's comprehensive school. Aunt Lal would describe the action and Mam would comment on it.

Aunt Lal: *'Bluddy 'ell . . . it's Ray Fletcher. Ay's wearin' a wig. Ay's got lipstick on!'*

Mam: *'Ray . . . Ray. D' yuh want me t' gerrit aht f' yuh?'*

If nobody laughed, she'd up her game a bit.

'Ray . . . Ray. Way always sedd y' were a puff. A' yuh goona gerrit aht forruz? Let's ha' á lukk.'

My dad never bothered with the Carnival, but Mr Walters was soon sniffing around. I think he'd been a miner at some point, but he was now one of the few men on the estate who'd opted for the 'Club' – on benefits because he was too ill to work. Men on the 'Club' were a different breed to men like my dad. But those differences showed themselves in subtle ways. Men like my dad would say hello and be outwardly courteous to men on the 'Club', but they would rarely sit next to each other on the bus or share a pint. Men like my dad would try to avoid being seen in the post office when men on the 'Club' were inside collecting their weekly giros. My dad would never lend Mr Walters his best hedge shears . . . and that had nothing to do with what may or may not have been happening with my mam.

Mr Walters had one suit and seemed to wear it all the

time. It may have been black at some point but was now stained and faded a shade or two lighter. He had a fine head of hair with touches of grey and a rough, roguish side parting. And no teeth, which allowed him to close his mouth more than most people. When he wasn't talking, he'd fold his bottom lip right up over the top lip so that it touched his nose.

Him and my mam made a formidable team. The chortling, leg-pulling and coarse humour went into overdrive, taken to a surreal level.

Mam: *'Wot's Mick Dink wearin'?'*

Mr Walters: *'Ay's gorra wumman's 'at on. Bigg 'un!'*

Mam: *'Wot y' gorrunder y' at, Mick? Let's ha' a lukk! Ah can gi' yuh a 'at. Dinky donkey! Shall ah cumm wi' yuh t' lavvy? They rekkon yuh've gorra bigg 'un!'*

Then they'd both laugh. And laugh. And laugh. My mam would pull out the handkerchief that she kept tucked into the waist of her skirt and theatrically dab her eyes.

'Ooh, ah'm goona wet me-senn in a minnit. Bluddy 'ell . . . oooooh, ah'll wet me-senn.'

Yes, I was embarrassed, but I also wondered if I needed to step in and . . . save her. Save her from herself. Watching my mam at the Carnival reminded me of the people I'd seen on the *Six O'Clock News*, chasing that wheel of cheese down Cooper's Hill near Gloucester. They started running and they couldn't stop. At the bottom, they tumbled and smashed into each other, often breaking legs or arms. My mam was currently halfway down the hill and gathering momentum.

I thought long and hard about saving her from herself but, in the end, I couldn't be bothered. So I walked off, had a ride on Mr Alf Morton's tank, bought a book on archery from the Salvation Army stall and went home to watch the wrestling. Tibor Szakacs vs Rocky Wall.

Following that first summer on the estate, I began my first year at Bagthorpe Primary School. The school was, as the name suggests, in Bagthorpe but, technically, the village of

Bagthorpe was still within Selston's boundaries and less than a mile away from my new front door. There was a special bus to take us to Bagthorpe every morning, but I often walked because I liked walking and it was always much more interesting. Yes, the buses from the Pit Houses had allowed me to glimpse my world with fresh eyes, but this new bus wasn't the same. It was a school bus, full of schoolchildren. Screaming. It offered n'er a second's peace to flick through the pages of a *Fantastic Four* or *Spider-Man* comic. No chance to daydream as we sailed past the White Lion and down Middlebrook Hill. No need for me to feign interest in the recent trade figures or pretend I was going to work. Rustle-rustle-fold-fold.

On foot, I had time to daydream. Time to make up all sorts of stupid shit. Each morning, Inkerman Road became a raging river, usually the Amazon or the Volga. The houses became abandoned spaceships. The hedges, out-of-control wildfires. Mr Tune's chickens, a pack of marauding Dromaeosaurid. I even asked Mr Tune if the chickens had ever tried to eat his dog.

He told me to '*Bugger off*'. But he smiled when he said it.

On foot, I noticed so much more. Out of nowhere one morning, on the jitty leading off the estate, I saw a goldcrest. According to my *Hamlyn Guide to Birds of Britain and Europe*, the goldcrest was very common, almost a garden bird, but I'd never seen one before. Chaffinches, goldfinches, green-finches, bullfinches, thrushes, wrens and jays were all regular visitors, but this was something new. Inside Mrs Wood's hawthorn hedge, little flicks left and right. I stopped dead in my tracks. Didn't even turn my head, just swivelled my eyes ever so slowly. There he was. A fabulous combination of colours: pale, leafy green and off-white, black flashes and that bright yellow cap. My new favourite bird.

I also saw my first barn owl on one of those walks. Ghostly and silent, it glided along the line of a rickety fence in the second field down from Hanstubbin, then turned towards me, passing barely 2 feet from my head. So close that I felt a faint ripple of air against my left cheek. Its dark eyes not seeming to notice me . . . just floating about, looking for stuff. My new favourite bird.

A more direct walk to Bagthorpe would have meant staying on the road, so I detoured up Inkerman Road, then down the fields that brought me out by The Shepherds Rest pub in Bagthorpe. I would leave home earlier than I needed to, giving me the chance to wander off the path and over towards the herd of cows in the top field. At first, they weren't sure what to make of me, but I persevered. I would sit on the grass, offering handfuls of dandelion leaves. They were good listeners, too. I would tell them about the books I'd been reading or some new technique I'd learned in maths.

'We call the number of marbles in a pile a triangular pyramidal number.'

As we now had a telly, I gave them a rundown of the week's highlights: John Laurie, the bloke from *Dad's Army*, reading *The Princess and Curdie* on *Jackanory*; Bronowski, of course. And the bands that I'd seen on *Top of the Pops* . . . Barry Blue and Alvin Stardust. Alvin sang a song called 'My Coo Ca Choo', dressed all in black with a big black quiff. He even wore black gloves and held the microphone like he was about to punch someone in the face. I demonstrated this to the cows, but without Alvin's quiff and all-black outfit, I couldn't really carry it off.

On the hottest days, I would wander across the dry grass, looking for crispy cowpats, lifting them carefully in the hope of seeing a few tunnelling dung beetles at work. A few years later, during the summer of 1976, the cowpats became so crisp and firm they could be skimmed across the fields like Frisbees. Summer was also the time when I would look out for early ripening ears of corn which could be stashed safely in my pockets and nibbled throughout the day like starchy Tic Tacs.

Information continued to stick to me like those viscous bits of cowpat I sometimes found under the crispy top layer.

Maths, in particular, was a constant source of joy. I loved its reliability. Do this, this and this and you get this. Like building Meccano. Or making toast under the grill. You just had to follow a few basic rules and everything was fine. Maths never sprung any surprises on me. I always knew where I stood with numbers.

All new schools feel unfamiliar and Bagthorpe was no different. It was larger than Selston Church of England Infant School, it had more classes and more people from more places. Some came from as far away as Brinsley and Eastwood. One girl came from Kimberley, which was dangerously close to Nottingham! And not so many of these new children had parents who were miners. Some were accountants and builders and policemen and owned fishing tackle shops. When I proudly announced that my dad was a miner, nobody seemed that impressed. Other lads made jokes about miners. Word had also got out that my mam and dad were much older than most other mams and dads.

Obviously, I let it be known that they weren't my 'real' mam and dad, but that didn't stop the jokes. About miners. About my dad's grey hair. About my mam's pants. One lad who'd been to our house with his mam, who was the clubby representative on the estate, told the class about my mam's underwear. Having grown up with them hanging from our washing line, it had never occurred to me that my mam's pants were particularly large or old-fashioned. At Bagthorpe, I learned that other mams bought Bikini Briefs, Saucy G-String Briefs and Lightweight Support Briefs, all of which came in various shades of blue, lemon, pink and purple. My mam had neither the inclination nor the figure to accommodate anything that was lightweight or brief. Her roomy, elasticated bloomers were heavily worn and vaguely whitish.

While not as physically painful as Sadistic-Bus-Stop and his twisty-twisty bollocks, the large underwear undoubtedly did more damage. The kind of damage that couldn't be moderated by learning to spell pterodactyl or dancing to

Dave and Ansell Collins. If I was going to make a stand, I needed something more . . . direct.

So, I signed up for a judo class in Jacksdale, run by a man called Nokky Hancock.

In the early '70s, martial arts was a thing. Even in Selston. Friday night on ITV, we all watched David Carradine starring as Kwai Chang Caine in *Kung Fu*. In the newspapers, we read about the success of *Enter the Dragon* and the mysterious death of Bruce Lee. We read about Angela Mao, aka 'Lady Kung-Fu'. At Mr Parkin's newsagent's shop, I was on constant lookout for *Kung-Fu Monthly*, a glossy magazine that unfolded into a massive poster of the sport's superstars. At the library, I found a book called *Bruce Tegner's Complete Book of Judo* and then, on the wall in the chippy, I saw a small flyer about Nokky's judo classes.

I wasn't bad either. As I'd filled out a bit, the basics, like break-falls and hip-throws, seemed to come naturally. Nokky called me a 'solid little bugger'. I started analysing stability and the position of my feet, and worked out a way of standing that was rock solid. In the playground at Bagthorpe, I'd let older kids run full pelt into my back, trying, without success, to knock me over. In my bedroom, I would assume that stance for fifteen or twenty minutes at a time, closing my eyes and letting all my weight sink down into my feet and into the floorboards. Bending my knees, minute bounces of the hips, deep breaths, like Nokky had showed me.

As it happened, the judo came in useful because Bagthorpe was home to a brand-new bully: Mark 'Grovey' Musgrove. He was a year above me and considerably bigger than me. Bigger than every other lad in the school. He also lived halfway down the estate, which meant I had to walk past his house every morning. Grovey's anger wasn't solely directed at me. He had it in for everybody. Lads who were good at football, girls who had long hair, younger kids who refused to hand over their white chocolate mice and Black Jack chews. Give him his due, Grovey was an equal opportunities bully, a galumphing giant with poor teeth and a reliably conservative approach to violence.

His first attack came out of nowhere. Eschewing the callous flair exhibited by Sadistic-Bus-Stop, he simply waited for his victims in the pub car park opposite the chippy, then marched forward and punched them in the stomach. In films on our new telly, people would just say 'Oof' then carry on fighting after they'd been punched in the stomach. Didn't work like that in real life, though. After the 'Oof', you dropped to your knees, fell forward onto your face and badly gashed your forehead. The inability to breathe was shocking at first and you'd spend a minute or two thinking you were going to die, but little bubbles of air eventually seeped into your lungs and the pain dropped to a level that at least gave you a chance to start crying.

For Grovey's second attack, I drew on my three or four weeks of judo training and prepared for battle. Not that it made any difference. Back on the floor, wheezing and grunting.

Third attack . . . same as before.

This needed newfangled ideas and fresh thinking. Instead of walking past Grovey's road in the morning, I cut across the fields at the end of our road, walked up to the White Lion, then across more fields to the path that led to Bagthorpe. Once safely at school, I kept tabs on Grovey, watching him during break times and at lunch. Watching him lollop around the playground, pushing people over and punching them in the stomach. I spotted an opening. After he punched anyone in the stomach and they fell down, he would bend over slightly, pushing his face into theirs, telling them how much trouble they'd be in if they told the teachers.

Mine was a simple plan. Anticipate the attack. Get close. Wait for him to strike and bend over. Smash my left hand into the back of his neck. A judo chop!

Worked well. Better than I thought, actually. He fell to the floor screaming and writhing in pain.

'Yuh've brokk me nekk!'

He added that he couldn't feel one of his arms. I can't remember which one. Dinner ladies were called, teachers were called, an ambulance was called. My explanation that he was punching a little lad in the stomach when it happened was, surprisingly, accepted.

The image of a young martial arts master was further enhanced when I picked up a black eye at judo. A real beauty. So good that I continued to whack myself in the face for several days, hoping to keep the eye closed and the bruising at its best.

I managed to make that black eye last all through Christmas . . . our first Christmas in the new house. And the first Christmas that felt like Christmas. Perhaps I'd been too young to appreciate the holiday season in the Pit House. Or was it because so few people came to visit – apart from not-aunt Aunts and grey-eyed thieves. Or because my dad stayed in bed for most of Christmas Day at the Pit House, leaving just me and my mam – a woman for whom the birth of Baby Jesus was little more than an opportunity to count up the number of people who hadn't sent us a Christmas card.

'*Way 'anna got one frumm Gadder-Legs,*' she'd sniff as I read out the names on the cards that came through our letter box.

'*An' wot's-'er-name an't sent owt. 'Er that used t' be married t' Cooky's lad.*'

And more.

'*Oositt . . . used t' wokk wi' Ken. Lottie's motther. Palled on wi' 'er wot died in that fire. An' can yuh remember Tommy Warren? 'An't 'eard frumm 'im. Ay can bugger off.*'

Every one of them. All the people that had ignored us.

But that first Christmas on the estate felt different. Bigger windows made the rooms brighter. And I could watch Laurel and Hardy, Basil Brush and a western. We got a green artificial tree from Woolworths in Alfreton and Dad put up a load of garlands. Aunt Lal and other women (and men) would pop round. Bottles of sherry were passed from glass to glass; records were played. Mam didn't need much encouragement and was soon up on her feet, dancing awkwardly and banging into things. This would make people laugh and they'd pour more sherry into her glass, filling it right to the top.

'*There's annuther one 'ere, Hilda.*' Then they'd laugh. and share sneaked, leery glances.

At some point, the St John Ambulance Brigade uniform would make an appearance. At the very same point, I tried snatching the sherry glass from a leery bloke's hand but he only laughed and leered some more.

'*Danny's tryin' t' stop yuh 'ayin' a drink, Hilda.*'

'*Dunna like may enjoyin' me-senn, duzz tha, lad? Ay's non mine, y' know.*'

'*Ay dunna lukk like yo,*' one of them commented.

'*'Is dad were Chinese or summat.*'

I couldn't hear what was said, but this revelation caused a few whispered asides. One woman smoked, cackled and pulled up the corners of her eyes.

'*Ah-so,*' she smoked and said. '*Ah-so. Arseholes.*'

I tried my best to feel sorry for my mam but knew it was a waste of time. I was sure that she wouldn't want me – or anyone else for that matter – feeling sorry for her. All of us and our pity could '*Bugger off*'. So, as Horst Jankowski's 'A Walk in the Black Forest' began blaring out of the Fidelity speakers, I buggered off to bed, wondering again about my mam and her blindness and why she never said anything about it and why she made such a fool of herself. And why Dad never said anything about it. And why nobody else said anything about it.

Despite the St John uniform, that first Christmas Day on the estate turned out to be surprisingly Christmassy. I'd acquired an alarm clock, which allowed me to get up before anyone else. My mam told me not to open any presents until she got up, but I opened mine anyway. People I didn't even know had bought me presents. Our new next-door neighbour. A family who lived up the top of our road. Books and socks and chocolate coins. On telly, a *Top of the Pops* Christmas special. Bing Crosby in *White Christmas*.

Mam came down the stairs in a good mood.

"Ave ah gorrenny presents,' she wanted to know. *'Gimme t' bigg 'unns fost?'*

The good mood soon dispersed – taking most of the spirit of Christmas morning with it – as she sat in her usual armchair and held out a hand.

'Weer a' they? Gimme 'em 'ere.'

I passed the presents to her, one by one. She opened them, as if she was in a film. An old black-and-white film. *Godzilla*. 1954. The scene where he attacks Tokyo. Instead of electrical pylons, her short arms tore apart wrapping paper and Sellotape. Instead of crumbling houses, her stomping legs were surrounded by discarded gifts. Instead of cannon fire, her cloud of smoke came from a freshly lit Player's No. 6.

In between puffs, I heard her version of his hellish, frenzied roar . . . *'Bluddy chocolates. Bluddy cardigan. Bluddy scarf. Ah've gorra ton already . . . dunna need annuther one.'* Any gift that wasn't immediately obvious was shoved my way. *'Wot's that?'*

'It's 'and cream. Frumm Vera.'

The scaly neck began to twist, pulling its head up and to the side. The grasping hands clawed the air. The mouth opened. Another drawn-out clang of impending doom. *'Dunna bluddy wear it. That can goo. Waste o' munney. Wot's that?'*

'Shortbread biscuits. In a tin. Frumm Edie an' fam'lee.'

'They tew sweet f' may. Mekk me teeth 'ot. Tin'll be oaraight. Putt summat in it.'

Even Dad got up. Old habits died hard, though, and he refused to join in with any present-opening ceremonies. The only gifts he ever received were tins of Old Holborn or packets of Jelly Beans, and they would stay unopened on the kitchen table for the whole of Christmas Day. And Boxing Day. Eventually, they'd be shifted, still unopened, to a drawer somewhere. And then? I never found out.

That first Christmas Day at the new house was also the

first time I'd ever had Christmas dinner at somebody else's house. Me and Mam walked up to Aunt Lal's. (Not Dad, obviously. He had no interest in eating his dinner at someone else's house. Did Mam miss him? Not really. Did I miss him? Even if he had come with us, I would have felt uncomfortable, watching him sit with all these people he didn't really know, wondering what to say. He was much happier in his kitchen, eating his peanut butter/sardine sandwich and listening to his records.) There was sherry and Watneys Party Seven. Already there were some of Aunt Lal's relatives, some small children, older children and a woman I'd never seen before pulling a chicken out of the oven. She paused only to light a fag, then brought the chicken into the back room, showcasing her talent for being able to inhale and puff smoke without using her hands.

Foolishly, I'd taken a book with me, Tolkien's *The Hobbit*, which Mr Brindley, one of the teachers, had been reading to us at school. A man I didn't know picked it up and started reading in a funny voice. I wasn't really listening, but it made everybody laugh. Especially my mam. Laughing. Too loud.

As I looked back at my dinner, I was seized with religious fervour. Or something very like religious fervour. In a formal voice and to the whole room, I explained that, at school, we always said grace before dinner. Every day, without fail. There were a few mumbles, but religion wasn't to be messed with – especially on Christmas Day. Waiting for silence, I launched into the usual 'For what we are about to receive', seamlessly segueing into the beautifully sombre lines of my favourite carol, 'In the Bleak Midwinter'. All five verses. Said, not sung. Cherubim and seraphim thronging the air. A heartfelt plea for everyone to join in the final 'Giiiivve myyy heart'. And finishing off with an 'Aaaaahhhhhhmmmmmennnnnnnn' that broke the twenty-second barrier.

Even with fag ash on the chicken, it was the best Christmas dinner I'd ever had. Watching them all mumbling along. A couple of children older than me, a couple younger. Young adults, older adults. All of them wondering if I was taking the piss. But it's religion. We'd better not interrupt him, just in case God is watching. And their faces. So unsure. The best Christmas dinner and the best Christmas present I'd ever had. All rolled into one glorious holiday whole.

I was so happy with my 'religious conversion' that, in the New Year, I announced that I was going to Sunday school at the Methodist chapel on Portland Road. The miners and Methodism got on well. The miners even had their own brand of Methodism called Primitive Methodism.

'The Primitive Methodist Church, so closely allied with the mining community, ought to adequately cater for the soul growth and the moral health of these workers. We are called to strike into the current of their daily life, and bring to their hearts and homes the health-giving Christ gospel with its lofty ethical code. Our business is to kindle warm spiritual fires, to reveal the ringing joy note in Christian life, to give keen edge to spiritual appetite and to foster romance in mining Church life,' said the Reverend Henry Fletcher in 1921.

Primitive Methodism for all of us Primitive People.

At Sunday school, I learned about right and wrong, morals and parables, played lots of table tennis and read bits of the Bible. Some of the passages I knew, which impressed the ladies in charge. More prizes . . . more books. For being a good reader and singer of hymns. For answering questions and asking questions.

Once more, I settled into the rhythm and tone of Bible-speak.

'For this is the will of my Father, that everyone who looks on the Son and believes in Him should have eternal life, and I will raise him up on the last day.'

'That I may know Him and the power of His resurrection, and may share His sufferings, becoming like Him in His death.'

And so on.

Was my 'religious conversion' genuine? Or was Sunday school simply an ill-advised attempt to join a gang? To belong? A bit of both, really. I knew all about the links between

Methodism and mining, and being at the chapel – unfussy, red-brick, built in 1899 – offered hints of how things used to be. Musty halls, families in their Sunday best, moustaches trimmed, boots shined, hats, white collars, bustles and babes in arms.

There were bits of furniture and door handles and pictures that had been in the chapel since it was first opened. I would stand next to them; looking, touching, smelling the last seventy-odd years. Despite the discomfort and menace that came with having Mam as a mam, those seventy-odd years gave me a counteracting comfort and reassurance.

The Selston of 1974 wasn't an excessively God-fearing community, but the chapel and its door handles had value . . . they carried some weight. And when I held one of those door handles or sang 'Be Still And Know That I Am God', I could feel my own body stretching and squashing itself – just a 64th of an inch, here and there – into the space that Selston had made for me many years ago. And, once settled, I saw Selston in all its . . . glory. That's the only word for it. Not religious glory. This was a spiritual glory, a human glory. All people that in Selston do dwell, sing to the Lord with cheerful voice.

Sadly, my Sunday mornings at the chapel didn't last. Going out on a limb, I decided to bring up the Blind Man of Bethsaida. My 'glorious' epiphany had set me thinking about life and what have you, so I asked one of the nice ladies if, like Jesus, the Devil could restore sight to a blind man. Or woman.

On no account, said the nice lady, was I to put my faith in the Devil. Her worried look let me know that I was now a marked man. Well, a marked nine-year-old.

Bizarrely, it wasn't the Devil that finally got me ousted from Sunday school. It was the dinosaurs. Although the Methodists weren't opposed to science, the nice ladies didn't appreciate me banging on about evolution. I got the feeling that . . . I knew too much stuff. Too many dates and Latin names.

The nice ladies in charge were all right about it, though.

They kicked me out, politely. I was even allowed to keep the books and prizes.

But not the table tennis bat. A shame, because I was getting quite good.

A Victory for Selston

We welcomed 1974 with another miners' strike, more black-outs and a three-day week. Once again, my dad was at home when he wasn't normally at home. Once again, I heard my mam moaning about having to cut down on fags. Although industrial action – the miners and other unions going head-to-head with the government – felt like it was becoming the norm, I wasn't particularly worried and naturally assumed that, at some point, it would all get 'sorted'. After all, this was mining we were talking about. The unbreakable bond between Selston, coal and all that history.

So I relaxed and made the most of it. Sitting with my dad at the kitchen table; him reading the paper, me reading my little *Commando* comic books. More candlelit dinners. More fun and games. On the estate, children enjoyed the darkness, fog, wind and mild temperatures. On the Top Rec, we played scrappy football matches in the darkness, fog, wind and mild temperatures.

On the radio and the telly, politicians and men who knew about these things argued about these things. Some called the miners 'a unique case' and 'a special case'. We were told that miners were the engine of Britain's economy. A bloke called Lord Wilberforce said,

'Other occupations have their dangers and inconveniences, but we know of none in which there is such a combination of danger, health hazard, discomfort in working conditions, social inconvenience and community isolation.'

That sounded about right.

Some, on the other hand, called the miners 'greedy little men' and 'bully boys'. They talked of militants, extremists and incalculable damage to Britain's industries. They hinted at 'an enemy within' and most fingers were pointed at the NUM.

No matter how vivid my imagination, it could never conjure up my dad and Mr Gregger Kirk as 'bully boys'. But the NUM was in militant mood and, as more unions came out on strike with the miners, the Conservative prime minister, Edward Heath, decided to ask not just my dad, Mr Gregger Kirk and Selston, but the whole of the UK: who runs this country?

In February 1974, he called a snap general election, expecting the voters to side with him against the miners and the unions. The voters didn't side with him and, in March, Harold Wilson became the leader of a minority Labour government that gave the miners a 32 per cent pay rise!

It had all been sorted.

On the radio and the telly, those same politicians and men who knew about these things hailed the general election as a victory for the working man. A victory for coal and the miners. A victory for us.

So, why didn't it feel like a victory? Why weren't people out in the streets of Selston, waving flags and kissing passers-by? Although Selston – part of the Ashfield constituency in a general election – had voted Labour since 1918, working-class support for the party and the unions was on the wane. People were getting fed up with the interminable strikes, go-slows, walkouts.

Knowing that my dad would be unwilling to talk about this shift in attitudes, I went to the library. In fact, over the next few weeks, I took up residence in the library, armed with my favourite pencil – a burgundy Eagle Graduate HB – and a red Silvine exercise book to make notes. With the help of Rosemary, my favourite librarian, I read lots of articles in lots of newspapers and magazines. I didn't understand all that I read, but I understood enough. It was obvious that Labour

– generally regarded as the party of the working man; the party that supported the miners – had won the election by the tiniest of margins. Admittedly, Heath had lost his battle with the miners and the unions . . . but only just. His 297 seats to Harold Wilson's 301 left an embattled Labour Party well short of an overall majority in parliament.

Even after the election, stories about the Labour Party – Tony Benn, Michael Foot and their mates in the union movement – were increasingly coloured by talk of the left wing and a socialist utopia. As if this was what Dad and the miners were fighting for. Socialist utopia? Neither Dad nor most of the people on the White City Estate had the slightest interest in a socialist anything. And if we wanted to glimpse utopia, all we had to do was look out of the window. There it was: Selston.

Admittedly, it was a somewhat muddled-up, paradoxical utopia. A utopia that had spent 700 years covered in coal dust. A thick layer on every doorstep, shopfront, ploughed field, school classroom and pavement. It was in the food we ate, the beer we drank and air we breathed. Being part of a mining family in a small, out-of-the-way village like Selston meant we were different. Just like that bloke, Lord Wilberforce, had said, mining families were assaulted by a 'combination of danger, health hazard, discomfort in working conditions, social inconvenience and community isolation'.

At the library, I read books that told surreal stories about miners who'd lost legs underground, banding together to beat up policemen with their wooden limbs. A miner who'd lost his sight but continued to challenge anyone he met to a fight. (He was known as the 'Blind Brawler' and I suspected that he and my mam would have got on well.) I found tactless passages that described miners as 'more like working animals than human beings' and 'primitive outcasts' (who no doubt deserved a suitably Primitive Methodism). I found passages that castigated the miners for their excessive drinking and fighting and immorality. I found passages that detailed the Aberfan disaster in 1966, the year after I was born: the collapse of a pit slag heap that had killed almost 150 people, most of them children attending Pantglas Junior School, which was completely consumed by coal-coloured slurry.

Through the library window, I could just about see up the slight hill to the White Lion. The other side of that hill was much steeper and ended at Bagthorpe Primary School. How long would it take a collapsing pit slag heap to ooze down that hill? Would it instantly level the red bricks of the small village school, crushing me and the other children? Or would it simply fill the spaces in between those bricks with mineral-rich quicksand? A graceful black wave squirting down the main corridor, plucking colourful pictures and typed notices from the walls and bursting through classroom doors? A slow motion suffocation? Small ink- and paint-stained hands reaching for the strong arms and reliable smile of Miss Hargreaves?

Did these things really happen to people like me? And Collo? And that lad on Hooley Street?

Rosemary found me a very, very old book called *The Condition and Treatment of the Children Employed in the Mines and Collieries of the United Kingdom*, which was based on an 1842 government report. Inside were sections that mentioned the Bagthorpe and Portland pits. I was in Selston Library actually reading about Selston in 1842. Yes, these things did really happen to lads in Selston.

No.170. Bradley Mart, colliery clerk. [at Portland]: They have 29 boys under 13 years of age and 27 under 18. The youngest is seven years old. They are let down and drawn up by a rope and no regard is paid to the number. They work from six to eight and are allowed two hours for meals when the engine stops. Three-quarter days are from six to four with two hours for meals and half days from six to twelve with no time for meals.

At No.2 [colliery] Charles Kirk, about a month ago, had his leg broken by the coal falling and is now out of work. Others have been hurt and a boy of 12 years old was killed by the roof and coal falling. At No.4 [colliery] Christopher Cresswell about 16 months since had his leg broken by the roof falling and was unable to work for six months.

The length of the list of British mining disasters surprised me. I totted up some of those deathly numbers with my Eagle Graduate HB. Not all of the numbers. Just a few. Over 100 men lost at Wallsend Colliery, Tyne and Wear, 1835; 50 at Arley Mine, near Wigan, 1853; 89 at Arley Mine, near Wigan, 1854; 361 at Oaks Pit, Barnsley, 1866; 294 at Albion Colliery, Glamorgan, 1894; 209 at Blantyre Colliery, Lanarkshire, 1877; 270 at Prince of Wales Colliery, Monmouthshire, 1878; 189 at Haydock Colliery, Lancashire, 1878; 69 at Trimdon Grange Colliery, County Durham, 1882.

And 18 at Markham Colliery, Derbyshire, July 1973. Markham Colliery. Near Staveley, where Great-Great-Great-Grandad Thomas had worked. July 1973. The previous summer; the summer we moved to the estate. Less than a year before the general election. Less than a year before the 'victory'.

I remembered Markham. Markham made headlines. In the *Daily Mirror*, it was called: 'The Cage of Sudden Death'. Markham was talked about in Selston's pubs and bus shelters, in shops and kitchens. Even in our kitchen.

The conversation was usually started by my mam when she heard the disaster mentioned on the radio. *'Worr abaht them blokes, eh?'*

Dad would try explaining details of the accident, but was quickly silenced.

'Ooh, shurrup, willya.' At this point, Mam would sweep her hand across the table, searching for her fags. *'Bluddy gooin' on abaht it. Ah dunna want t' know.'*

I wanted to know. So I asked my dad. *'Why din't t' cage stop?'*

'Shurrup!' More sweeping for the fags. *'Ah'm sick o' 'earin' it. Bluddy ding-dongin'.'*

Despite my mam's best efforts, those thoughts and worries were always there. In every house on the estate. Every day Dad and his mates went to work, every time we heard an ambulance, every time we saw the funeral cortege follow a coffin down Holly Hill Road. A reminder that mining was a filthy, dangerous job – filthier and more dangerous than any other civilian job you could think of.

If the story of mining had ended there, with just the filth and danger, Dad and the men from Selston would no doubt have been questioning their sanity. Catching the bus to the Labour Exchange, looking for jobs that weren't out to murder them. But mining came with a . . . quirk. This filthy, dangerous job gave Dad and his mates something the Labour Exchange rarely had on its books: a sense of pride, purpose and camaraderie. Dad and his mates shared life and death. Fear and pain. Fags and sandwiches. Dad and at least five generations of his family – my adopted family – had lived and died providing fuel for the Industrial Revolution, the electrical revolution, the rail revolution, the motoring revolution, all those factories that had equipped British forces in two world wars and Harold Wilson's 'white heat' technological revolution of the 1960s.

Mining allowed Dad to stand shoulder to shoulder with all that history and say, *'Ah were theer. Ah played me part.'* Mining made sure that Dad mattered. All we asked in return was that Dad and his mates were given a decent wage and the council charged us decent rent for a decent house. With the basics sorted, our not particularly socialist utopia would soon fall into place: we ate well enough, I enjoyed going to school, the milk and papers got delivered every morning, Carl Douglas was at number one with 'Kung Fu Fighting' and we had a telly.

As if to back up my argument, I headed home with my notes, waiting for Dad to wake up and come downstairs. I waited till he'd made his first cup of tea, then brought up the union bigwigs and asked him, point-blank, if he had any plans to overthrow Britain's capitalist economy.

'Nott this wikk.' He shook his head as he unscrewed the lid from the peanut butter. *'Ah sedd ah'd cut Cookie's 'edge forrim.'* Then he added. *'Way doin' oaraight. May, thay and thee motther.'*

Yes, we were doing all right. Some of the gobby young

miners being interviewed on telly were complaining about not being able to afford a Morris Marina, but Dad didn't take any notice. He'd bought a new drill instead, replacing the old D500. Mam was able to smoke loads of fags. She even bought some new clothes and started going to a pub in Somercotes with Aunt Lal, usually on a Wednesday night. Dad put some relatively expensive – over £1 a roll – intricately patterned wallpaper on the landing wall. He started doing the football pools. Mam bought a radio for the kitchen. I got a Raleigh Chopper for a belated birthday present – just over £1 a week for thirty-eight weeks from the clubby.

Indeed, life seemed to be chock-full of 'good' things. Dad took me to the pictures for the first time in my life, to see *Mosquito Squadron*. He even came with me on a school trip to the RSPB Lodge in Sandy, Bedfordshire.

We saw a pied wagtail. It reminded me of the lapwings.

To find out what the working man was really interested in during the turbulent years of the mid-1970s, all the union bigwigs had to do was come to Selston. I would have shown them Dad's new drill, given them a tour around our estate, taken them to the post office on pension day, sat with them on the bench opposite the chippy, waited outside while they supped a pint in the Bull & Butcher and finished off by letting them buy me a Lolly Gobble Choc Bomb from Wilde's Corner. No revolution, no class war and not even the slightest attempt to overturn the capitalist economy.

Sadly, none of those union bigwigs took me up on the offer. If they had done . . . the next decade might have turned out very differently.

Waiting for the Avon Lady

Dad had many talents. One of them was the ability to break wind as he descended every one of our thirteen stairs. Always finishing off with a mighty flourish . . . parp-parp-parp-parp-parp-parp-parp-parp-parp-parp-parp-parp-paaaaarp! Dad and me laughed at that. He showed me cartoons in the paper and we would laugh at the latest exploits of the *Daily Mirror*'s downtrodden working man, Andy Capp.

We laughed about my mam, too. About her being blind. Well, we didn't laugh about the actual disability. Dad just had . . . a bit of fun. And that made us laugh. Little things. Moving the salt and vinegar off the kitchen table, just after Mam had put them on there. Then he'd watch, smile and wink at me as she sat down with her dinner and struggled to find them. Or he'd shut doors that were normally kept open. Then he'd watch, smile and wink at me as she walked into them. Or he'd swap a couple of her fags for some foul-smelling, super-strength Polish fags that he'd got off his Polish mate at work. Then he'd watch, smile and wink at me as she nearly choked to death.

Had Dad always played these mischievous games? Was he only choosing to share his secret now that I was older? Dad wasn't a vindictive man but he seemed to enjoy these 'bits of fun'. And he began encouraging me to have 'a bit of fun' at this blind woman's expense. So I would move the Andrex from its usual place on the floor by the Domestos.

Or mix up her shoes, putting odd ones together.

Then watch, smile and wink.

I was beginning to take a keen interest in Mam's blindness. Mam was a tiny woman and I was a tall boy, which meant I could look directly into her strange eyes. The white bits were duller than my white bits. The circular bits in the middle – a definite shiny green iris and black pupil in my eyes – were all misshapen and mangled in hers. More of an unfinished grey splodge.

I wondered how many other people knew she was blind. Like me, some of the people in the Pit Houses must have known or guessed that she was blind, but after we moved to the estate, Mam kept it a closely guarded secret. She never talked about it with me or Dad. Dad never talked about it with her or me. It was never mentioned. Not even in passing. How had it happened? When had it happened? Was she already married to Dad when it happened? Did it happen when she was a child? Was it an accident? Was it foul play? Was it hereditary?

For a while, after we moved to the estate and she started talking about people and things she'd 'seen', I believed her. Maybe she had come to some new agreement with the darts king. New eyes for a second-hand soul or something like that.

It didn't take long for me to work out that there'd been no new agreement. No new eyes. Mam had changed, though. Something or someone was driving her on, willing her to keep the blindness under wraps. Her deceits were cunningly pre-planned and full of clever details. Whenever a casual acquaintance drifted into earshot, she'd make some miraculously astute comment about next door's garden gate or the new bus stop down by the petrol station. She'd mention the afternoon's brightening sky or the design of the new label on a Co-op baked beans tin.

Mam, I soon realised, was an excellent listener. She heard people mention things, then, in turn, she mentioned them. Seemingly made in passing, these off-the-cuff comments were carefully designed to convince the world that her eyes were A-OK.

It annoyed me more than I thought it would. Annoyed me enough to watch and wait for the Avon Lady. As she approached our gate, I quickly ran round and drew all the downstairs curtains. Mam always talked about the weather with visitors (having heard the weather report on the radio) and that day was no exception. She stood before a pair of drawn, dark brown curtains, telling the Avon Lady that it *'Lukks a bit cloudy.'*

I watched. And smiled. And winked.

After a few weeks of being taken down to the shops by Aunt Lal, Mam decided to enlist me as her outdoor guide and accompanied me to Alfreton. As we set off for the bus stop, she held me tightly by her side, then began steaming ahead with big, bold strides. Down our road, round the corner, down the hill, down the jitty. To the world, it looked as if she was just another mother taking her young son to Alfreton, but if you were close enough, you could hear Mam's constant stream of questions.

'A' way near t' steps? Is there any dog mukk? Ah can smell grass. Is summboddy mowin' t' lawn? Wot's that noise?'

I would provide a running commentary about the world and everything that was happening in it. *'Five steps cummin' upp. No dog mukk. Fred Jeffries is mowin' 'is lawn. Summboddy drillin'.'*

'A' tha raight, Fred? Garden's lukkin' nice,' she'd shout.

On the bus, my job was to quietly tell her if anyone she knew was already on there. *'Minnie Routledge on this side.'* And I would tap the relevant arm.

'A' tha raight, Minnie? 'Ow's Derek?'

As we moved to the back of the bus, I also told her that Mr Bowmer had just come down the stairs and was getting off. He wasn't. I had no idea where Mr Bowmer was. But I watched, smiled and winked as she waved and shouted down the bus.

'A' tha raight, Cliff? Weer y' bin?'

I knew. Knew that I was being cruel. Why was I being cruel? Because I thought my mam should tell Selston and the world that she could not see. That she should face the truth. That she should admit she was blind. Unable to see the new bus stop. Unable to see if it was 'a bit cloudy'.

Being a Clever Bugger, I had sat down and worked out the pros and cons of Mam's life. Worked it all out with my Clever Bugger logic and reason. Worked out that everything would be so much easier for all of us if she simply told the truth. In a place like Selston, the truth had no desire to hurt or hoodwink. Just look what happened when the truth came to visit; when other people knew (or guessed) that my mam was blind. Those people made a big fuss. They treated her with care, gave her leeway and made her cups of tea.

'Duh yuh want sugar, Hilda?'

Dad, despite his 'bit of fun', did his best to deal with reality. He would always help Mam sort out her wardrobe, telling her which blouse was blue and which one had the nice pattern on it. He even bought a sewing machine and learned how to repair Mam's coats and put new zips in her slacks. Not many miners in the 1970s would have done that.

If she got that kind of personal service from truth, why not shout that truth from the rooftops? Why not wear a big badge or have a T-shirt printed? 'I'm blind! Now, make me a cup of tea, you bastard!' Why all the subterfuge? All the planning and preparation. All the listening. All the off-the-cuff comments. All the lies.

Lesser mortals might have given up and gone for the truth/badge/T-shirt option. Lesser mortals might have gone for logic and reason. But not my mam. She strode forward and stomped all over logic and reason, giving them a kick in the teeth for good measure. Regularly testing her mettle, she painted the kitchen (including bits of the cooker and most of the sink). She trimmed (butchered) hedges and bushes in the garden. She continued to cut (butcher) my hair and her own. If she and Aunt Lal bumped into someone at the shop, Mam would comment on what was in their basket . . . after

Aunt Lal had quietly revealed the contents of the basket from the far side of the shop.

'*Ooh, yuh've gott that noo gravy pahder,*' my mam remarked. '*Is it oaraight?*'

Obviously, Aunt Lal must have known she was blind. But what about the person buying the gravy powder? With careful planning and casually keeping her head turned to one side, had Mam really managed to fool the world? Or was the world simply playing along with this game? Being polite and not mentioning the obvious.

Worryingly, my mam continued to invite people into the house without knowing who they were or why they were knocking on our door. Greeting them like much-missed friends. Even more worryingly, those people continued to steal things.

And the old favourite, of course . . . setting fire to herself and almost burning the house down.

Like the Pit Houses, this new house had an open coal fire. Despite not being able to see what she was doing, Mam had managed to cope at the Pit House and the morning blaze would generally spring into life with little more than a Zip firelighter, a few bits of kindling and half a bucket of coal. But the chimney of the new fire was due for a clean and getting things started often required extreme measures.

After lighting the Zip firelighter, kindling and coal, Mam would attempt to 'draw' the fire by holding a sheet of news-paper across the opening of the fireplace. The trick was to leave a small gap at the bottom, allowing fresh air to rush in and fuel the nascent flames. Sometimes, the sheet of news-paper would catch light and the sudden whoosh would cause her to let go and push it into the fireplace, where it would be lugged up the chimney and released high into the sky. If her timing was off, the newspaper would float up to the front room ceiling, dangerously close to another of the 1970s' fire hazards: polystyrene ceiling tiles.

It was the smell that gave it away. Sharp, choking and

sulphurous. Heady and intoxicating, with a sweet but unbearably bitter tang. The scent of the ages. As my mam came out of the front room and back to the kitchen, I smelled then saw a few puffs of blue-black smoke. Blue, like my dad's suit. Black, like Elvis' hair.

Slowly, I opened the front room door and looked inside. The sharp scent of the ages stung my eyes as the blue-black smoke made good its escape, sucked backwards up the chimney in a hurried, shapeshifting column. On the ceiling, a couple of the tiles were melted and dripping, and the floor was covered in orange ash and small scorch marks. I smothered them with a cushion. Then poured water on them. Then opened the windows.

There were several similar incendiary incidents in the mid-1970s and not long after one particularly close shave, I came across a newspaper story about Hiroo Onada. Back in 1944, Onada, a Japanese soldier, had taken up his post on a Pacific island and then carried on fighting for thirty years. Even when the police and local fishermen came to tell him the war was over, he refused to listen. Indeed, he continued to wage a guerilla war, burning the islanders' food supplies and killing several of the poor souls who tried to explain that Japan had surrendered.

Onada was no simpleton. On the contrary, he was a graduate of Japan's top training centre for intelligence personnel and people encouraged him to run for parliament after he finally returned to Japan. Reading his story, I couldn't help but think of my mam. And I couldn't help but smile. In many ways, Mam and Hiroo Onada were kindred spirits. They had both fought on against all odds; against logic and reason. They had defied reality and refused to confront what appeared to be an obvious truth.

When compared to this fifty-two-year-old multiple murderer who'd fought the Second World War for three whole decades, Mam's way of life was no longer . . . unfathomable. Far from it. Like Hiroo, she was simply soldiering on. Unlike Hiroo, though, she never shot people or blew them apart with hand grenades; she just painted the kitchen, talked about the weather and told people she didn't need any help with the Sunday dinner. What's wrong with that?

That was when it dawned on me. Mam wasn't an exception, she wasn't alone. I cast my mind around Selston and began to recognise others who soldiered on. Others who doggedly defied reality and fought their unwinnable war in our muddled-up utopia. Top of the list was The Texan, a bloke slightly older than my dad who was always dressed in cowboy boots, jeans, brown leather chaps, black shirt, black leather waistcoat, bootlace tie and brown ten-gallon hat. His hefty gun belt weighed down by two impressive silver six-guns (real guns!) and his shiny spurs clink-clink-clinking as he sauntered past the Co-op. He sometimes drove an old Hillman Imp, too, but you never really got the full effect of the spurs unless he was changing gear.

The first time I ever saw Tex – as he was known by his friends – I was with my dad. It was an early Saturday evening and we'd been to collect fish 'n' chips. He was leaning against the bus stop, hat drawn low enough to hide his eyes, but high enough to show a grizzled chin and a barely smouldering cigar.

As we passed him, Dad nodded. *'A' tha raight, Tex?'*

Tex nodded back, silent but kinda friendly, readying himself for the ten-to-six stagecoach to Bandera Falls, via Underwood Miners Welfare.

A cowboy. Living in Selston. A real cowboy.

''Oo were that?' I asked my dad breathlessly.

Dad looked down at me, his face earnest and respectful. *'That . . . is The Texan.'*

The Texan's war wasn't the war of some flash dandy or preening peacock, dressing to impress the drab Selston world. His war was neither ostentatious nor garlanded with pretention. It was neither cause for laughter nor mockery. Tex lived a life that was in no way more unusual than any other in Selston. Tex was remarkably . . . unremarkable. The only difference was that he had at some point decided that Selston, circa 1975, was Fort Worth, circa 1881. And his unremarkable life would be lived beneath a ten-gallon hat.

There was also the Smoking Man. A man who used to march around the village. Smoking. Rage-fuelled, incessant smoking to match his rage-fuelled, incessant marching. Marching towards his rage-fuelled, incessant war. Always marching in the middle of the road.

This was no problem on our estate, but there were a couple of main roads in Selston that led to Nottingham and Mansfield. While not exactly rammed with traffic, there were enough cars and lorries to make them dangerous if you were walking on the white line for any length of time.

But, like Tex, this man's actions and his war were regarded as perfectly natural. Like Tex, he was greeted with nods and friendly enquiries.

'A' tha raight? Weer tha off tew?'

'Ah'm gooin' t' Wilde's. Gett summ fags,' he'd explain as he hurried on by.

Once or twice, concerned onlookers would offer a kindly word of advice. *'Y' ought t' gerr on t' corsey, me duck. Yuh'll get nokked ovver one o' these days.'*

Did he get 'nokked ovver'? Of course not. In Selston, he was safe.

Children were in on the act, too. Children like The Stingo, a lad who matter-of-factly insisted that he couldn't feel pain. At school, he would ask other kids to stab him in the hand or arm with a brass compass that he carried around for just such occasions. Although misery and discomfort were written across his face, The Stingo never flinched or yelped or turned down the offer of a mid-morning puncture wound.

One girl from the estate up by the church used to ride to school on a St Bernard dog; presumably because the family couldn't afford a pony. An older lad, Stanno Relph, would only play football if he had his dad's pipe clamped between his teeth. He was good enough to be picked for the school team but refused to relinquish the pipe and the referee wouldn't let him play.

To an outsider, these may have seemed nothing more than amusingly rustic affectations. The foibles and peculiarities of mining-village life in the 1970s. But those outsiders didn't understand. They didn't understand Selston and its power. The power to hold reality at arm's length. The power to win an unwinnable war. Triumph against all odds.

Over the centuries, something had happened to this tiny, out-of-the-way village. Something had been unearthed. Alongside the coal and the clay, Selston was full of . . . magic.

And this magic came in all shapes and sizes. Like parents and shoes.

I am not blind,' said the magic.

'I can fly.'

'I am impervious to traffic.'

'I smoke a pipe because it helps me score goals.'

'I helped the Devil repair his car.'

'I do not feel pain.'

'I am a personal acquaintance of Wyatt Earp and Doc Holliday.'

Even though I was only a child, this idea of 'triumph and magic' made perfect sense to me and my muddled-up utopia. It made sense to my logic and reason. It helped me understand my mam and The Texan. It helped me understand Stanno Relph, The Stingo and . . . me. It explained my life. The life of Selston and its people.

Well, almost. Something was beginning to niggle. There was a detail that didn't feel right and proper. That line from *The Condition and Treatment of the Children Employed in the Mines and Collieries of the United Kingdom*: 'a boy of 12 years old was killed by the roof and coal falling.'

I was ten years old in 1975, just two years off my twelfth birthday. I wondered how I'd feel if those two years were all I'd got left. And how quickly you die when a roof and coal falls on you. It seemed an unfair way for a twelve-year-old to die. Much better to come a cropper on your Raleigh Chopper, trying to break the three-minute barrier for a full circuit of the estate. Or getting your head lopped off by a particularly sharp cowpat Frisbee. That's how twelve-year-olds should die . . . not crushed and smothered 300 yards from sunshine. If Selston was full of so much triumph and magic – which I genuinely believed it was – perhaps we ought to set aside a bit more for the miners.

As I'd been locked out of Sunday school, I wondered if my niggle would make more headway with the darts king. If he could restore sight (albeit briefly) and win all those darts matches, surely he could manage the long-term eradication of dermatitis and a cure for coal dust.

So off I went to the Welfare in Jacksdale, where he'd been made captain of the darts team. I asked if anyone had seen him. A woman behind the bar raised her head and shook it, ruefully. He'd been in a couple of weeks back, she explained, but had told everyone some cockamamie story about an elderly relative in Spalding. And then he'd run off with the charity jar.

'*Ovver twenty-six quid,*' she told me, soft tears welling in her hard eyes. Soft tears rolling past puckered lips that sucked hard on a John Player Special king size – the blue box that boasted of middle-tar with the full-tar taste. Soft tears that reflected the hard, gaudy colours of the Railroad Fruit Machine. The woman looked across at the machine's flashing lights. '*An' ay wunn t' jackpot on theer. Annuther tenner.*'

For several weeks, I was plagued by odd sensations. The hand that had once shook the Devil's hand seemed to burn and prickle. I saw his Elvis hair dancing before me. In my ears, a peal of cruel laughter. And in my nostrils, the sulphurous scent of the ages. They would reveal themselves as I was in bed or walking to school. When I was watching telly or playing darts. Even when I was in the House of Trees.

At those moments, I called out. I read all I could about the Vauxhall Viva and offered various bits of advice about oil changes and an upgraded exhaust.

But there was no answer.

And no favour.

We Came Across a Young Blackbird

No matter what else was happening in Selston, none of it seemed to matter when Dad and me were in the garden. And as I watched him, delicately stepping from gooseberry bush to lilac tree to foot-long runner beans, I often wondered if he'd constructed some sort of giant force field/greenhouse that kept him and the Brussels tops safe from all that other stuff . . . out there.

Our back garden on the estate was much bigger than the one we'd had at the Pit House: longer, twice as wide and bordered at the bottom by a young oak tree and a couple of mature silver birches. Local birds were obviously allowed inside the invisible geodesic dome and it wasn't long before Dad and me had a line of robins, blackbirds and starlings following the turn of our spades. The starlings sometimes came mob-handed and would always retreat to the left-hand birch tree, which was obviously some kind of meeting place for the estate's starling population. The biggest gatherings always took place on crisp, late autumn evenings, with a clear sky and just enough light to see all the flashing starling colours and shapes among the mustard-mottled birch leaves. As the birds began to gather and chat, the commotion became loud enough to carry from the back garden, through a window, through a wall and into the front room where I'd be watching whichever programme filled the melancholy slot that marked the end of children's weekday telly.

There, rootling about underneath the theme tune to

Roobarb or *The Wombles* was a mix of tweets, whistles, trills, descants, croaks, rattling melodies, gliding hoots, machine-gun chirrups and robotic bleeps. Although starlings never had much of a reputation in the birdwatching world, there was an understated, iridescent brilliance to their green, purple, gold and black plumage that was vividly complemented by their unconventional song. Pure and almost lyrical, yet feverish and other-worldly.

If the birds were loud enough to drown out *Roobarb*, I knew it was worth going upstairs to my mam's bedroom window and pulling back the net curtain. It might take twenty minutes or twenty seconds, but one bird would suddenly dart out of the birch leaves. A murmuration. Like an explosion of tumbling dominoes, the rest followed (only 200 or so, but an impressive sight nonetheless), tipping from the tree and heading vaguely towards the bottom of the estate. After a few seconds, there was a sudden change of shape and the black dots would corkscrew round and round and left and right, squeezing themselves together before stretching out like a giant rubber band that twanged over towards the school playing fields and back past the Parish Hall, to be finally flung up the estate and out of my sight. Then round again. And again. No circuit ever the same.

Despite all the houses and all the people, this bit of Selston was home to a large and exotic(ish) collection of wildlife. As well as the goldcrest and the barn owl, other regular visitors included hedgehogs and grass snakes, foxes and badgers, kestrels and sparrowhawks, weasels and stoats, hobbies and redstarts. Down at Moorgreen Reservoir, there were shrikes, herons and divers.

Unfortunately, this bit of Selston also included half a dozen genuinely troubled lads with an unhealthy interest in all that wildlife. The kind of lads that had distant eyes. The kind that shot at cows' udders with air rifles. Or tied fireworks to cats. Or fed live bubs (baby birds) to ferrets.

One lad – in my year but much smaller; his fair hair always shaved into a tight skinhead that made him look bald – had this thing about collecting animals, especially birds. He used to put various traps in hedgerows around the village, then keep whatever he captured in a shed. His reputation had amassed a small group of 'followers', a mixed bunch who glided around the school playground at lunchtime, calling themselves The Flying Lizards. A congress of scruffy little reptiles whose main pastime was attacking girls and the occasional dinner lady. I did join them – briefly – relieved to find a posse of excitable young boys who were as deluded about gravity as I was. We gathered in a small wood attached to Bagthorpe school, known as the Plantation, where we ran around, climbed trees, jumped across ditches and made lizard noises.

Bagthorpe Plantation had an odd shape. A truncated crucifix or a verdant mushroom cloud, it stretched from Bagthorpe to Underwood, half a mile long by a quarter of a mile wide. When Mr Brindley read *The Hobbit*, I immediately thought of the Plantation as Bilbo and his chums entered the black forest of Mirkwood. Like Mirkwood, the Plantation's trees were dense enough to block out the sunlight and tall enough to make you think there could be something nasty living at the top.

Near the centre was a fallen oak that served as both our headquarters and a makeshift climbing frame. Even jumping from its highest point of 7 or 8 feet rarely caused any injuries. The Plantation floor was covered with years' worth of leaves, mulch, grass, twigs, moss and soft earth; it was like landing on a deep, comfy mattress.

As we rarely saw anyone else there, the occasional sound of cracking twigs or swishing branches caused us to be on the alert. We always hid if we heard anyone approaching because we didn't want grown-ups gatecrashing our party. We'd run over to the far end of the Plantation, away from the school, where there were several overcrowded patches of bracken. So thick and squashed together that you could easily imagine walking right across the fine carpet of curled leaves.

Most of the time, we would simply crash into the edges

of the bracken patch and crouch on our haunches till the sound of other people disappeared, but I once crawled right into the middle of the largest patch. After a few silent minutes, I heard the others head back to the oak tree. I didn't. Slowly, I spread myself on the ground, easing the tough, sinewy bracken stems aside until I could lie flat on my back. Thanks to my dark woolly hat and camouflage jacket, I felt unseen. Transported to a world that existed just beneath the surface of our own. A genuine Jack o' the Green.

All around me, I began to notice spiders, beetles, ants, slugs, frogs and bracken bugs. I could hear something scuttling past my head and just caught the sight of a hedgehog doing his rounds. It didn't take long for me to become part of the furniture. Bugs and ants crawling across my jacket, something cold on my hand. I was glad I wore my hat.

Slowly, I inched my way back into the open, shuffling along on my back. As my head popped out of the bracken, I was struck by the sudden lack of movement. Out in the Plantation, everything was still. Not even a wood pigeon or a crow clodhopping through the trees. The wood pigeons, crows, blackbirds, blue tits, magpies, insects, reptiles and hedgehogs in the Plantation knew better than to draw attention to themselves when a bunch of young boys was on the prowl.

We did once come across some wildlife. A young blackbird, frightened and hunched up under some nettles. We made our lizard noises, but the bird didn't really move. It just hopped a bit. Another blackbird started flying around us, screeching and swirling about our heads. The mother. Somebody threw a piece of wood. It clipped the young bird, which started flapping its wings but still didn't move anywhere. The mother screeched louder still, swinging frantically across the canopy. I'd never heard an animal panic before. It sounded almost human. The sound any mother would make if a group of sadistic idiots was throwing stuff at her helpless child.

Eventually, one of the other lads walked up to the bird and prodded it with his foot. The young blackbird fell over and flapped its wings some more. It also began to make little squeaky noises and I saw some white pooh splurge out of its backside. Nobody said anything.

The bald skinhead whooped and drew back his right leg. Thunk! The bird spun wildly through the air and rolled over towards me. Its head was lolling over to the side and its dark, beady eyes drilled deep into mine. How much do animals know? How much did that juvenile blackbird understand? It knew it was going to die, I was sure of that.

And I was right.

The poor bird was still alive as he began tearing off its wings and legs. I can still see its face. I was standing just to the left, no more than 3 or 4 feet away. The bird was screeching and screeching, even as the lad was trying to pull out its tongue. He couldn't manage it, so he threw the bird to the ground and slowly squashed it with his boot. After he was finally satisfied that it was dead, he pulled down his trousers and had a shit on the bloody remains. He kept calling the bird a bastard, while the shit was coiling into little piles on top of it.

That was the last time I flew with The Lizards. For several weeks, the images and noises of that afternoon kept landing in my head with a thump. As if thrown from a great height. The pathetic little stick of a leg detached from the writhing body. The slow ooze of blood and quick twitch of flesh. The piercing shriek of life being ripped apart . . . of a child realising it was no longer in its mother's care. A fear that was far more destructive than any physical pain.

It was not something I wanted to be part of. Not only because of what they were up to, but also because it was glaringly obvious that I didn't really enjoy being 'part of' any kind of group or gang. Three or four people was about my limit. Any more than that and the whole thing became unwieldy . . . difficult to control. Unless it was a game of football.

I didn't explain why I was leaving The Lizards and I didn't expect any black marks against my good name as a result of this departure. A couple of the younger lads seemed to have lost heart, too. They didn't mind running around the playground attacking dinner ladies, but, like me, they spent less time at the Plantation.

As soon as I decided to go solo, I began taking an interest in the big green hill that sat behind the houses on the other side of our road. Why hadn't I thought of that before? It was perfect, marking both the top boundary of our estate and the highest point in Selston. What with the gale-force winds that regularly aimed themselves at its south-facing crest, I would have no need of The Lizards or the trees in the Plantation. All I had to do was stand there in my parka and wait. The wind would do the rest, taking me on incredible airborne journeys across lakes, oceans, mountains, valleys . . . and, as I was a bit older, up and down the M1.

Although Selston was always classed as 'rural' or 'out of the way', we were dangerously close to the M1. Just a few hundred yards of flat grazing fields separated the motorway from the north-east border of the village, somewhere between Junction 27 and 28.

I would roam north to Junction 29 and Chesterfield, home of the famous Crooked Spire. South to Junction 21 and Leicester Forest East Services. Dad and me had visited Leicester Forest East on that trip to the RSPB Lodge and I was fascinated by the idea that you could eat egg, chips and beans while sitting above six lanes of traffic. Leicester Forest East looked like the future. Like the world would look in the space-age twenty-first century.

I did most of my flying in the hours of twilight and darkness. Less chance of being spotted. At night, up above the motorway, I was always struck by how splendid it looked. Long, meandering lines of concrete lamp posts, each one crowned with a modern orange sodium lamp. Running through the middle was an ever-modifying matrix of headlights, tail lights and indicators that chased themselves up and down the country. The motorway had a soundtrack, too. The hard rumble of the lorries or the howl of a high-tech Japanese motorbike. Irate horns and the speedy flap-flap of a partly deflated tyre.

Sometimes, a single car caught my eye and I would dive down to see who or what was inside. Usually, it was a man on his own; listening to the radio or cracking open a window to dispose of an empty fag packet. There were families, too. They interested me the most. A whole family, sitting in a car, zooming towards the huge hazy glow of Nottingham and Junction 26.

Unlike Nottingham, Selston didn't have much of a 'glow'. Apart from a few splashes of orange sodium on the main roads, the rest of the village was still lit by ordinary 100-watt household bulbs hanging from rusty green cast-iron lamp posts. And most of those bulbs had been smashed by miskicked footballs. From a few hundred feet in the air, much of after-hours Selston was invisible. The pits were easy to pick out, and the pubs, of course, but there was little left of our estate save the sporadic flicker of a bathroom or kitchen light.

Those night-times held a darkness that felt full of wonder. Even when I wasn't airborne, I need only lie on the little lawn at the back of the house to marvel at the colour of Mars, the waxing and waning phases of Venus, the Orion Nebula, the occasional moon of Jupiter and the translucent splurge of the Milky Way.

Most of our large back garden was dedicated to runner beans, peas, cabbages, nobby greens, strawberries, guzgogs, rhubarb, taters, carrots, lettuce, parsnips, blackberries and blackcurrants. I had my own set of tools and my trusty mini Qualcast lawnmower. Admittedly, Dad handled most of the digging, but I was head weeder, head waterer and chief of pest control. Birds were kept away by fine netting. Slugs, snails and caterpillars (mainly on the cabbages) were removed by hand and relocated to the compost heap. Any that I missed were hoovered up by visiting hedgehogs. Teamwork.

The 'record-breaking' summer of 1976 really was record-breaking. Everyone on the estate knew this because the weathermen on the telly drolly acknowledged each of the fifteen consecutive days when the temperature reached 32 degrees. Watering the garden during a hosepipe-banning heatwave wasn't easy, but Dad and me came up with an elegant solution. Instead of washing with soap and shampoo, I simply splashed about in my twice-weekly bathwater and wiped myself down with a flannel. I would then take the hosepipe up to the bathroom, drop one end into the water and feed the other end out of the window. Back downstairs, I would suck on the hosepipe, starting the flow of water from the bath, which gravity then transferred to a couple of old copper boilers. I did occasionally wazz in the bathwater, but Dad insisted that it was natural and wouldn't hurt the carrots.

Admittedly, the lack of soap did make me a bit pongy towards the end of August, but it was the 1970s. Everybody was a bit pongy. The whole world was a bit pongy.

Although Dad grew his own taters, he wasn't averse to a few freebies. During tater-picking season Dad and me would set off early on Sunday morning and fill a couple of carrier bags from a field up by the church. The apple and pear scrumping involved a fair bit of climbing over fences and up trees, so Dad left that to me.

One other source of food was the Dumbles, several square miles of farming fields, paddocks, meadows and overgrown thickets that stretched from Selston to Westwood and Bagthorpe. Access was via a mysterious cinder path that lay close to yet more of Selston's soot-coloured back-to-back terraced houses. Why was it a mysterious cinder path? Because it ran from nowhere in particular and ended somewhere else that was nowhere in particular.

The tarmac-covered, modern bit of Selston petered out 10 or 20 yards from the start of the path, as if the workmen had realised that was as far as they were going to get with their lorries and road-making machines. The result was a slight tear in the fabric of time. On one side, it was the mid-1970s. There were cars, Give Way signs and double yellow lines, and the papers were full of stories about the IRA, the three-day week and Tom Baker becoming the new Doctor Who. But if you looked the other way, towards the cinder path, you caught a brief glimpse of a world that was 50 or 60 years old. Car-free, hand-washed laundry, hobnail boots by the door and three generations under one back-to-back terraced roof.

Dad was born less than a quarter of a mile from that very spot – a small row of small houses at the top of Buxtons Hill – and he'd been taking the same walk down the Dumbles for most of his life. He knew the people that lived in the houses and the people who owned the farms. And he knew every inch of the Dumbles. On weekend walks, he would show me where to find wild strawberries, the biggest nettle patch (to make fertiliser and nettle beer) or a nest or two of wood-pigeon eggs (fried or boiled). He knew the best time to pick goosegrass for his broth and dandelion leaves for his tinned pilchard sandwiches. He collected rosehips for rosehip syrup, elderflowers for elderflower tea and could even find blackberries in winter. He pointed out the difference between hemlock (poisonous) and cow parsley (tastes a bit like carrot). He showed me which mushrooms you could eat and which ones might kill you.

We often took Dad's air rifle with us, the one I once thought would be used to dispose of my mam. Obviously, Dad had either got cold feet or forgotten all about it – I suppose there was also the slight chance that I may have misinterpreted him collecting the rifle from a mate he'd lent it to as uxoricidal intent – but these days it was only used to shoot at cardboard targets nailed to the trees by Bagthorpe Brook. (Dad didn't mind taking eggs from nests, but he didn't like the idea of killing an animal for fun.) Nothing much happened on those weekend walks. Apart from the metallic rattle of Dad rolling fags on his Old Holborn tin and the snap of the rifle, we sat in comfortable silence for as long as that silence lasted. Neither of us wore watches, but we both knew when it was time to go home. Dad slipped the rifle back into its case, then over his shoulder, and we collected that day's fruit 'n' veg bounty.

We took a different route on the way home from the Dumbles, following the brook until it hit the road at The Shepherds Rest pub. If we got there before last orders at ten to three, Dad would have a bottle of Manns Brown Ale, with a glass of shandy for me. Then we'd trudge the half-mile or so to the Dixies Arms, turn left and walk up the fields that took us home.

Even when Dad let me carry the rifle, nobody stopped us or said anything. Seeing a young boy with an air rifle or a shotgun was commonplace. There was actually a bizarre, brief period when some lads and I held air-rifle battles in Bagthorpe Plantation. Eight or nine of us would split in to two teams and then start shooting. If you got hit, you were 'out'. Not to mention severely bruised and scarred for life, depending on the type of lead pellets we were using.

I was once shot in the neck, one lad had to have a pellet removed from his calf and another lost an eye. He made the best of his bad lot, though. Popping the glass eye out during lessons, frightening girls with it or leaving it on the teacher's desk.

Walking from the Dixies Arms, up the fields to Hanstubbin, was a wonderful way to finish the day. The path rose as we headed towards Selston and I would turn around every few steps, just to see how much more of the backdrop had been revealed. Bagthorpe Village and the church at Underwood. The cluster of trees that stretched away to the left. The soft, rolling hills on the right. And that vast East Midlands sky, all blazing seashell-blue except for the flat-bottomed white clouds that inched their way towards Jacksdale and Brinsley. Even the giant pylons, stabbed into the earth at various points, couldn't spoil the view. Faced with so much nature, the pylons became softer, friendlier, cartoonish . . . almost charming.

You saw that a lot in Selston. You saw it a lot in the whole of the East Midlands. Heavy industry – metal, brick, concrete and electrical cables – plonked beside bucolic postcard views. For some, a coal mine set amongst fertile meadows would be a walloping great blot on the landscape, but when you see the two of them side by side every day of your life, the boundaries begin to blur. You notice how the hard, geometric building shapes have effortlessly slotted in with the randomness of the surrounding fields and hills. The way the sunlight blasts off the canteen windows at certain times of the day.

I should really have asked my dad all sorts of questions when we were on those walks. Can you remember the first time you went down the pit? Have you ever been scared? But I didn't because . . . well, we didn't really say much at all on those walks. I got the sense that he preferred not having to answer my questions, so I never asked him any. I didn't mind the lack of conversation. A change from being with my mam, constantly explaining what the world was up to and why.

There wasn't complete silence, we did talk about some stuff. Goosegrass and rosehips. The unevenness of the fields. A particular patch of the hawthorn hedge where a single branch had grown much higher than the rest. We talked about how dusty the air was during hay-baling season. And we always did a slight detour just before we got back onto the road at Hanstubbin so we could sit on a large gate while Dad quickly smoked a quick fag. The gate was at the entrance to a meadow, 20 or 30 feet wide, with a tall, straight hedge that ran down either side. It was no more or less beautiful than any of the other meadows in and around Selston, but those green walls harboured strange enchantments. As we sat there, I felt as if I could hear the grass humming. A single, mysterious oscillating note. The kind I'd heard on the many sci-fi/supernatural children's programmes that stuffed the 1970s telly schedules: *Doctor Who*, *Children of the Stones*, *The Tomorrow People*, *King of the Castle*, *The Owl Service*, *The Stone Tape*.

Was it the humming of small children from centuries past? Children who'd sat on the same gate and played their games on the same grass? Had my dad been one of them? Or my grandad? Was that Grandad Jack's voice I could hear, echoing all the way from 1890-something?

Although I hadn't had much input into our family's history, all the stuff I'd learned about Selston and coal had given me a hint of something that I was part of whether I liked it or not. Great-Great-Uncle Enoch, Great-Great-Great-Grandad Thomas, Crich Stand and the Bevin Boys. And although this 'gang' had grown to many more than three or four people, it didn't seem to matter. I still felt welcome. It was nice to know about the people and the years that came before me. Balancing on the top of that gate, I could actually see the house where Dad was born. Was that why he always stopped at this same spot? To reach out`and . . . touch.

One of the books I'd recently borrowed from the library was *The Gauntlet* by Ronald Welch. Written in 1951, it tells the story of a young boy who finds a rusted medieval gauntlet. When he slips the gauntlet on, he is transported back to 1326. After reading it, I felt sure that the gate had acquired similar powers. If I jumped off the gate and into the meadow, would I land in 1890-something? Playing Deer Stalker or Stick and Hoop with my grandad. Spinning my handmade thaumatrope. Wondering if the future really would look like Leicester Forest East Service Station.

The Tin Hat

I walked a lot. Enough to know every inch of Selston and its neighbouring villages. From Pye Bridge to Pinxton to Kirkby-in-Ashfield to Annesley Woodhouse and all points in between. Selston began to feel huge. As big as the world. As big as I would ever want the world to be. Yes, there were other places I might visit every now and then – Alfreton, Mansfield, Sutton-in-Ashfield, Nottingham – but they existed outside of and strikingly separate to my world.

And, like any world, Selston was home to different types of language (abrupt and broad on the estates; weirdly forced and posh up by the church), landscapes (everything from rolling hills and dense forests to flat agricultural fields and the craggy edges of the Peak District), flora and fauna (hardy gorse, elderberry and bracken on the old pit heaps; magnolia trees and colour-matched shrubs in the detached non-council houses).

As I lived on a council estate, I was automatically marked out as a particular kind of person at school and in daily life. The fact that I was a Clever Bugger made not the slightest difference. The fact that I could read an Ordnance Survey map didn't stop children and people who didn't live on the estate laughing at the state of my trousers. The fact that I was beginning to understand the connection between light speed and time travel didn't alter the fact our house was a bit grubby and there were odd curtains in the landing window – one blue, and one with white and burgundy stripes.

But, even on the estate, among all that grubbiness and the odd curtains, there were levels within levels. Wheels within wheels. Worlds within worlds. For instance, some houses on the estate didn't look like estate houses. They were immaculately presented, with a paved area for a second-hand car. Occasionally, a child from one of those houses would somehow become my friend and I was invited to his house in the school holidays to play Battleship or Action Man. My mam would always insist that I go because she wanted to know what his immaculately presented house was like inside.

''A they gorra phone?'
'Ah think so.'
''A they gorra freezer?'
'Ah think so.'
''A they gorrall them pictures on t' walls?'
'Wot pictures?'
'Pictures. In frames.'
'There's a picture wi' a man an' wumman wi' no clothes on. Standin' on a big swan.'
'Wot d' they need that on t' bluddy wall for? Bluddy swan!'

One house I went to had a leather sofa in the front room and the lad's mam would always cover it with a plastic sheet if I went to sit on it. Eventually, I realised that she didn't do this with everyone. Just me. That seemed fair enough. When I wasn't at school, I mostly wore jeans and they were mostly dirty – I figured out that I could easily wear the same pair of jeans for a month without having to wash them. But, at eleven years old, I was also old enough to know that the plastic sheet gave me special status. Estate status. Usher-me-straight-into-the-kitchen-if-they-had-company status. Like Sadistic-Bus-Stop's neck twisting, it could have caused me huge psychological pain . . . but it didn't. I started wearing my jeans for two months. I started rubbing extra bits of muck into the knees whenever I was due to visit the lad and his plastic-sheet mam. If being part of her gang meant washing my immaculately filthy jeans, I was happy as a lone wolf . . .

prowling around her living room, looking for a chance to deposit fluff and dust on her spotless three-seater.

I was also old enough to notice that plastic sheeting wasn't the only change on the estate. Some people on the street below us – a youngish couple with no children – had booked a holiday to that place: 'abroad'. They came back to a mixed reception. My mam and many others reeled off all the reasons she didn't want to go 'abroad' – quality of the toilet roll, smell of the fags, having to eat dog meat, too many foreigners, no teapots.

'*They dunna use teapots ovver theer. They just mekk it in . . . a cup.*'

Cars were also becoming, not exactly commonplace, but certainly less un-commonplace. Games of football were interrupted by angry men who shooed us to other parts of the street. One man used to cover his yellow Ford Escort with a large plastic sheet – much larger than the sofa sheeting – every night. The estate streets were not much wider than a car, so by the time four or five were parked on our road, football was unofficially banned.

Once again, my mam became suspicious and immediately stopped speaking to any member of any family that acquired a car. Unfortunately, the march of time, technology and hire purchase was against her. Like Canute, she could not stop the flood of Escorts, Allegros and Maxis and she all but gave up when one of the blokes Aunt Lal was seeing turned up in a Triumph Dolomite.

She even surprised me and Dad by suggesting that Dad could learn to drive. I watched his face and knew what he was thinking. He was wondering how he would read the paper if he was driving. How he would roll a fag. How he would chat to Pete Day if he was in a car and Pete was on the bus.

He finally shook his head. '*Ah'm non bothered.*'

While reading about dinosaurs, I'd come across the idea of evolution, but it had never occurred to me that evolution also happened to humans. Was I witnessing evolution in real time, right here on the White City Estate? Abroad, cars . . . pictures on walls?

Perhaps because my mam couldn't see them and Dad wasn't that bothered about them, we had never hung pictures on walls. But word had obviously reached my mam that this was the latest fad.

'*Summ on 'em ay 'em gooin' upp t' stairs,*' she scoffed. '*Yuh canna lukk at pictures when yuh gooin' upp an' dahn t' stairs.*'

As luck would have it, I was given a front row seat for Selston's evolutionary upheaval when I secured my first paper round. Seven mornings a week, I would be roaming to the outer limits of the village. Past Selston Church of England Infant School, past the almost invisible remains of the old Bull & Butcher Pit, past two Co-ops and two chapels, down a steep hill and up another steep hill to a sizeable chunk of non-council houses in an area known as Top 'n' Town.

As soon as she realised I would be in constant touch with this other world, my mam was full of questions about the houses and the people who lived there.

'*'A they gorra phone? 'A they bin abroad? 'A they gorra freezer? 'A they gorra car? 'A they gorrall them pictures on t' walls?*'

'*Ah dunna know. Ah just shove t' papers through t' letter box.*'

'*Yuh can 'ay a lukk through t' winnduh, can't yuh.*'

The following morning, after depositing a copy of the *Daily Express* through the letter box of the first non-council house, I walked the few steps that took me to the front window. I made a mental note of any ornaments and pictures on walls, holiday mementos, types of furniture and wallpaper, size of telly, etc, and was about to wander back down the yard when a woman in a dressing gown opened the front door.

'*Me mam wants t' know if yuh've gorrennny pictures on t' wall,*' I explained with my best paperboy grin.

She said something to me, but I honestly had no idea what she was talking about. That's what I meant by 'weirdly forced and posh'. This woman lived three-quarters of a mile from my house and I could not understand a word she said.

The rhythm of the words . . . the shape of her mouth as she said them, the look in her eyes. It was all wrong. Words with too much detail. Like words sitting on top of other words.

Unsurprisingly, it turned out she was complaining about me looking through her window and relayed that very same complaint to Mr Parkin, the newsagent. By the time I returned my fluorescent-yellow heavy plastic newspaper satchel, I had been moved to a different paper round. Without so many posh houses.

That didn't stop me walking about Selston, though. And the more I walked about, the more I noticed the subtle signs of change . . . tarmacked driveways, iron gates, houses with names like The Lodge, Golden Pond and Oak Tower. Conversely, the more I walked about, the more I understood Old Selston. The church, dating back to 1120. All those Methodist chapels, built in the early part of the nineteenth century. Wansley Hall in Bagthorpe, part of a medieval manor house that was as old as the church. But so much of that sense of history came from me knowing that most of Selston was built on the ruins of several old pits.

At night, I had visions of long-dead miners clawing their way through the soil surrounding my dad's carrots and parsnips. On quiet afternoons, I would put my ear to the back lawn, listening for the deep and distant clank of hammers and picks. Thunderstorms would become underground explosions; the opening of a new seam or something more sinister, like a build-up of deadly methane. In heavy rain, I would imagine ice-cold rivulets working their way through soil and rock; refreshing trickles that ran across the arms, hands, necks, backs and furrowed brows of the miners. The tiniest gesture that King Coal was on their side.

Whenever Dad and me were digging the garden, I was on constant lookout for any sign of Selston's past: time-worn coins, trinkets, pottery, rusted tools, snap tins and pit checks (brass tokens that were used to record how many men had gone down to the pit bottom). A reminder of Selston in its mining heyday.

Coal was and always had been Selston's destiny. From the early bell pits (bell-shaped holes, dug to a depth of 20 feet and more) to shallow shafts in the seventeenth century and shafts of over 1,000 feet in the nineteenth century, Selston had been mined by several companies, including Barber, Walker & Co, Kirkby Fenton, James Oakes and Butterley.

More companies and more mines in Selston meant even more people. Even more people needed even more places to live, and the mining companies began building rows and terraces of miner-friendly housing all across Selston. These working people needed somewhere to drink, hence Selston's excess of pubs. They needed somewhere to worship, hence the establishment of all those chapels throughout the parish.

But the most important building in 1970s Selston wasn't any of these. Not the collieries or the church or the chapels. Not the main Co-op or the Parish Hall. It wasn't even the school or the chip shop. It was the Ex-Service and Working Men's Club, a modern-ish single-storey rectangular red-brick building that everyone knew as the Tin Hat. (The name came from the fact that British soldiers, the ex-servicemen, wore tin helmets. The Memorable Order of Tin Hats is an ex-servicemen's organisation that actually began in South Africa after the Great War and quickly caught on in the UK.)

Entry to the Tin Hat was via a set of dark, wooden doors that led you through a tiny porch, then into the main room and its multicoloured vinyl tile floor; all restrained post-war reds, blues and yellows in repeating square patterns. To the left was a small stage, flanked by lavvy doors, and just to the side of the stage was a complicated-looking bingo machine, filled with numbered balls and adorned with various tubes, pipes and lights. To the right of the main doors, a bar ran for almost the whole width of the hall, and behind it a line of cramped shelves and hanging optics was ready to dispense the bizarre concoctions that most of Selston's women drank: whisky and pep (peppermint cordial), gin and orange (cordial), rum and blackcurrant (cordial), Cherry B tipped into half a pint of lager. Ask for a glass of wine and you'd be laughed out of the door, into the car park and all the way down Chapel Road.

For the men, there was beer. Bitter, mild and a few bottled stouts. That was it. And if a man didn't drink bitter or mild (or the bottled stouts), the other men in the Tin Hat would spend the entire night looking at him with narrow, suspicious eyes.

'Wot's tha game, yoth?'

Even young lads like me were allowed to have a crafty half. Men would wander past, pointing to my drink. *'Putts 'airs on y' chest, that duzz.'*

Yes, the men and women and children drank, but there was no 'booze culture' in the Tin Hat. Yes, the men were over the limit when they drove home in their Austin Allegros and Ford Cortinas, but nobody started fights while the bingo was on.

The Tin Hat was open all week, but Saturday was the big night. Every inch of the floor was covered with tables, chairs and stools – apart from a small area in front of the stage that was reserved for dancing. Each table in turn would be covered with drinks, packets of crisps, tins of snuff – J & H Wilson and McChrystal's – overflowing ashtrays and bingo tickets. And each chair or stool had at least one bum on it. If it was Christmas or New Year, you needed to arrive bang on opening time or there was little chance of getting a seat. We always arrived bang on opening time because my mam liked to sit at the same table, just a few feet from the ladies' lavvy door. That way, all she had to do was stand up, grab hold of the large cast-iron radiator and follow the wall to the door.

Dad came with us, but rarely sat at our table. Dad was on the Committee, a group of blokes – all miners – who looked after the Tin Hat's finances, booked the bands that played there, ran the bingo and the weekly raffles. The Committee had a special table in the middle of the floor, where they would sell bingo and raffle tickets and make any important announcements via the special Committee microphone.

'*Can the owner of a white Ford Escort, registration number ATL 404G, please move the bugger.*'

It was strange to hear these dour, dyed-in-the-wool East Midlands miners sounding like newsreaders on the telly whenever they spoke into the special Committee microphone. As if they realised that the electronics suddenly made their proclamations dazzlingly public. Words like 'of', 'the' and 'to' were pronounced in their entirety and there was a vague attempt to include aitches in 'home' and 'have'.

Unsurprisingly, some of the miners, including my dad, wanted no truck with the special Committee microphone and its devious electrickery. Those men continued to make announcements and read out winning bingo tickets in the old-fashioned way: hollered between pulls on a fag and sips of a pint.

Aunt Lal came with us, joining me and Mam at the table near the lavvy, their chair backs resting on the bulky metal radiator, laughing with or gossiping about anybody who caught Aunt Lal's eye. Not that it was easy to catch anyone's eye in there. By eight o'clock, the smoke – from fags, cigars, pipes and possibly a pile of burning tyres – would settle just below the ceiling like a thick black cumulonimbus. By nine o'clock, it would have expanded downwards to a level of about 2 feet. You could see it shifting and swirling about as people made their way to and from the bar. Great purls wafting back and forth whenever someone opened the double doors to the outside.

There were a handful of small fans attached to the walls, but nothing less than a typhoon would have completely cleared the Tin Hat. The fans would simply nudge the smog from one bit of the room to another. Windows were opened in the summer but, during the colder months, the only escape was in the lavvy (the scent of urinal blocks and Domestos felt weirdly refreshing) or the Committee Room.

What the Committee used it for was anyone's guess, but the Committee Room contained a large wooden table, several benches, some boxes and, most important of all, fresh air.

Access was via a door at the side of the bar and, over time, it became a sort of children's playroom. On busy nights, there might be twelve or fifteen of us in there, eating crisps, playing dominoes or talking about what had happened on that week's episode of *Voyage to the Bottom of the Sea*.

The only time we saw an adult in there was if we were making too much noise during the bingo. A snarling man's head would poke round the door and tell us to, '*Keep bluddy quiet.*' No matter whose head it was, we knew better than to disobey or argue. One lad, a bit older than most of us, once told the snarling head to '*Bugger off*', then laughed. The snarling head disappeared and was replaced by the lad's dad who crashed through the door, roared his dissatisfaction, grabbed the lad's shirt collar and hauled him back through the door in a matter of seconds. We never saw him again.

The poor lad had interrupted the bingo . . . and that had to be paid for. In blood. As soon as the room heard the immortal words, '*Eyes dahn forra full 'ahse*', the Tin Hat went deathly quiet. The fruit machines were switched off, no one went to the bar or lavvy and crying babies were taken outside and shot. The only sound was the occasional strike of a match or the squeak and scratch of markers and pens on the precious bingo tickets. The entire room was in thrall to the imperious voice of the bingo caller, reading out the numbers as they were propelled from the clear tube at the top of the ingenious bingo machine.

'*Downing Street . . .* (dramatic pause) *number ten. Kelly's eye . . .* (dramatic pause) *number one. Two little ducks . . .* (dramatic pause) *twenty-two. Five and six . . .* (dramatic pause) *fifty-six. Top of the shop . . .* (dramatic pause) *blind ninety. All the threes . . .* (dramatic pause) *thirty-three. Unlucky for some . . .* (dramatic pause) *thirteen. On its own . . .* (dramatic pause) *number four.*'

Although the grown-ups didn't talk during the bingo, they were allowed to shout out at various points. Phrases like '*Ah'm sweatin'!*' if somebody only needed one number for a win. Always a mass wolf whistle following the appearance of the number eleven. *'Legs-eleven* . . . (dramatic pause) *phweep-phweep.'* Or one of the time-honoured calls – *"Ere, 'ahsey-'ahsey,'* etc – that signalled a winning ticket. Immediately after a win, the garble and shash of voices would rise again to a cacophonic level as the whole room discussed drink orders or how close they were to taking the night's jackpot of 5 quid.

Us kids weren't the only ones who got told off during the bingo. Occasionally, a couple of new adult faces would appear in the Tin Hat. Out-of-towners. Unsure of the etiquette, they would try something foolish like going to the lavvy or ordering a pint in the middle of a game, thus shattering the intense concentration that gripped the room.

'Wot tha upp tew?' One of the Committee men would suddenly shoot up from the special table. *'Can tha wait till t' game's ovver?'*

The offender, confronted by a chorus of wrathful murmurs, would skulk back to his seat and light a fag, hands trembling as an immeasurable fear forced his struggling heart well above 250 beats per minute.

The Tin Hat took its bingo very seriously.

In between the winning tickets, there was music. Country-and-western duos, rock quartets, blokes who sang to backing tapes, pop bands, skiffle groups; all playing a collection of golden oldies and current hits. Certain bands were well known on the northern/Midlands club circuit and would draw huge crowds. I can remember one lot, much loved by The Texan, who played rowdy bluegrass tunes and Johnny Cash covers. And another bunch who dressed in glittery shirts, leather trousers and knee-length boots, belting out songs by Slade, Roxy Music, David Bowie, Sweet and T. Rex.

The Tin Hat crowd was not exactly 'with it', but the opening chords of 'The Jean Genie' or 'Cum on Feel the Noize' was enough to bring a few twenty- and thirty-something couples out of the woodwork. Young women in revealing, lurid-green blouses and revealing, canary-yellow skirts or revealing, wildly patterned tunic dresses.

Accompanying them were young, hard mining men with shoulder-length hair, sideburns, electric-blue three-piece suits and burgundy platform shoes.

You could see the generational dividing line, right there. Still seated, Selston's older women, with immaculately shampooed and set hair, and colourful but conservative blouses. Staring haughtily at the brazen, immature hussies. And Selston's older men, with sombre brown suits, trilbies and neatly trimmed hair. Shaking their heads as the frisky stallions thrust nimble hips to songs that sounded nothing like Jim Reeves.

These dancing, prancing men weren't scruffy. Their flash suits were clean, their showy shirts were ironed and their shimmering medallions were polished with Brasso, but they stood in direct sartorial opposition to their fathers and uncles. What was most confusing for the older men was the fact that these younger men still sounded and acted like men from Selston. They pulled their weight on the night shift, they fixed their own cars and gave their kids a clout when needed. But they did all these things while choosing to spend their Saturday nights dressed 'like bluddy wimmin".

Children were allowed to dance, too. Most of them just ran around and slid on the shiny vinyl-tiled floor, but my dancing came straight from the heart. Music meant something to me . . . I had learned to harness its power. Almost immediately, I was able to gauge and codify any new song I heard on the radio or the Fidelity record player. The songs soon became more familiar than the days of the week. More gratifying than the taste of my favourite sweets and chocolates. More loved and longed-for than the people I knew. The stomping buzz and wail of T. Rex's 'Metal Guru'; Roberta Flack's voice gliding across 'The First Time Ever I Saw Your Face'; the mysterious pub-piano-driven melodies of Lynsey de Paul's 'Sugar Me'; the pre-electronic groove of 'Popcorn' by Hot Butter; the epic Mediterranean balladry of Vicky Leandros' 'Come What May'. If a Tin Hat band played a song I knew, that song would compel me to pick up my glass of orangeade and position myself right in front of the stage.

I'd never been taught to dance, but I watched *Top of the Pops* every week and could manage a fair impression of the people I'd seen in the audience. My chubby little arms would match the singer's words with fluid jerks and gestures, while my joggling fingers would copy a tricky guitar solo or keyboard trill. Within a couple of songs, my head and T-shirt were soaked with sweat, and I could hear the occasional bark of encouragement from somewhere in the club.

'Goo on, Danny-lad!'

Eventually, I'd peel off my T-shirt, wipe down my chubby arms, chubby face and chubby belly, and recharge with a long swig of orangeade. Then I'd close my eyes and gulp a few dramatic deep breaths before my feet slid back into the rhythm. I would spin and point, fall backwards onto my hands like the older boys who danced to Northern soul, stomp my cheap slip-on shoes in random patterns on the square tiles. For a chubby kid in shabby trousers, I had a spectacular sense of rhythm, often dropping into half- or double-time movements. My shoulders and elbows shaking to some complex, sixteenth-note counter-rhythm that only existed in my head.

One night, a much older girl moved to my side and danced with me for the rest of the band's set. I had a sneaking suspicion that she felt sorry for me. Throwing myself around like a dying fish stranded on a sandbank. Probably thinking: 'Poor lad. They must've let him out for the weekend.'

I couldn't have cared less. All that mattered was the music and the rudimentary spotlights that blasted out shards of red and green from the tiny stage. And the applause. As I walked back to my mam's table, the singer would say, *Let's gi' 'im a big 'and,'* and the whole club clapped and cheered. Sometimes, the barman would bring over a free orangeade. The young men in electric-blue three-piece suits would shake my hand as they walked past, while their wives and girlfriends would ruffle my sweaty hair.

'*Yuh lukked stoopidd,*' my mam would laugh.

How did she know I looked stupid? Even with binoculars or a really powerful telescope, she wouldn't have been able to see me.

'*No, ay dances luvvly, don't yuh, me duck.*' Sometimes, the women on the next table would stick up for me, at which point my mam would take a long, red-faced swallow of her whisky and pep.

Anyone looking for the beating heart of Selston in the 1970s would have found it right there. In the Tin Hat. Distributed amongst those tables and chairs. Soaked into the beautifully designed beer mats and drunk from the battered tankards that hung behind the bar. Echoing from the neatly painted walls and the nicotine-stained ceiling. Written about in the strange language of scuff marks, blots of spilled beer, stiletto dents and Seg scratches that decorated the dance floor. You could see it in the Brylcreemed hair and the set of false teeth sitting on a windowsill (put there by some old bloke to stop them clinking on his pint glass). You could smell it in the heavy aroma of perfumes and talcs from Avon and Estée Lauder. You could hear it in the lyrics of 'Juke Box Jive', 'Gudbuy T'Jane', 'It Must Be Love', 'Mull of Kintyre', 'You Sexy Thing', 'Don't It Make My Brown Eyes Blue', 'Yellow River' and 'Kentucky Rain'.

There was something else, too. The mutual respect and understanding that existed between the men on the coalface was also common to the Tin Hat. Even though those men might be singing along to 'Tiger Feet' by Mud and cracking open another packet of Rothmans King Size or John Player Special, nothing that had happened underground was disregarded or forgotten.

See him in the waistcoat? He saved the life of that bloke standing at the bar, dug him out from a roof fall. The two sitting right at the back were waiting to start their shift at Markham when it happened. That big lad coming out of the lavvy carried the broken body of his own father from the coalface after he was killed by a rolling coal cart. Sometimes, men would arrive at the Tin Hat straight from a shift, as evidenced by their coal-dust eyeliner and vicious thirst. Their first pint was, of course, on the house.

From the outside, the Tin Hat looked like it was made out of red brick and mortar, but that was only a disguise. It was built from thousands of nuggets of coal and, just like the people inside, it was held together by trust and tragedy. Inside the Tin Hat, you became part of a communitarian ecosystem made up of mines, beer, music, hard graft, neat gardens, coal, stoicism, impending disaster, unpretentiousness and quiet pride.

Inside the Tin Hat, you became part of Selston.

Skegness

The Tin Hat even went on holiday en masse. On a spring Saturday night, a flyer would appear on the noticeboard by the bar, advertising a week's holiday in Skegness (known by everyone as Skeggy) or nearby Ingoldmells and Chapel St Leonards (known by everyone as Ingo'mills and Chapel) for July or August. By the end of the night, every holiday spot would have been filled. Perhaps two or three coaches of Selston families all going to the same caravan site for the same seven days. As if White City Estate had been temporarily relocated to the Lincolnshire coast.

It would start early on a Saturday morning, men and boys carrying cases and bags of food – in case they didn't do Typhoo tea or gold top milk on the east coast – up to the Tin Hat car park. Men from the Committee would load everything into the coaches' luggage compartments and help small children climb the steep steps into their luxurious interiors. Ashtrays for every seat, plus your own funny-shaped bit of silver metal for striking matches and stubbing out fags. A netting pocket for newspapers and comics; skylights and fresh air vents.

It was, I overheard Mrs Anston say, 'tip-top'.

As soon as the coaches left the Tin Hat car park, bottles of pop were popped, fags tapped from packets, card games started, shoes, boots, hats and false teeth removed. There was knitting, dominoes, apples and chocolate bars. I unpacked comics and books, slotting them into the net pocket in the order I planned to read them: *Warlord*, *The Victor*, *Twenty Thousand Leagues Under the Sea*. Knowing that today was the big day, a few Selstoners would line the route along the first bit of Nottingham Road, waving and laughing as we went past.

By the time we were rattling down Annesley Lane, one of the Committee men would take charge of the coach's microphone and PA.

'Nah then . . . probbly gooin' t' tekk us three o' four 'ours, dependin' on traffic. Sit back an' enjoy t' ride. 'As ennyboddy gorrenny matches? Ah left mine in t' sootcase. Ennyonnya enny-onnya?'

Several hands would shoot up from headrests, waving boxes of APD, England's Glory and Swan Vestas. One set of nicotine-stained fingers offered a massive box of Cook's Matches, over 200 in every box. At fifty a day, that box might last till Tuesday.

The coach always followed the same route. From Selston to Annesley, skirting past Newstead Abbey, ancestral home of Lord Byron (I recited the opening line of 'She Walks in Beauty' to Jonno E, the lad sitting next to me: 'She walks in beauty, like the night.' He barely raised his head from that week's issue of *Shoot!*), then down tree-lined roads that took you through Ravenshead and Nottinghamshire's so-called Hidden Valleys, on to the mining villages of Blidworth and Rainworth.

We were in the thick of what would have been Sherwood Forest and Robin Hood territory. The Major Oak (reputed to be Robin Hood's secret hideout) was just up the Old Rufford Road in Edwinstowe and there were claims that Blidworth was both the resting place of Will Scarlet (one of Robin's Merry Men) and the birthplace of Robin's bit on the side, Maid Marian.

After Rainworth, bigger roads would speed us to Newark and Lincoln, the huge cathedral and its three towers. Strange people on busy streets, watching as we inched through the holiday traffic. Tin Hat children waving at those strange

Lincoln children and strange Lincoln policemen. Surly Tin Hat teenagers waving two fingers at those strange Lincolnshire teenagers and strange Lincoln policemen. Then instantly regretting their folly as the coach ground to a halt at traffic lights and the policemen strode towards the coach doors. 'Phhsssstt' went the hydraulics as the doors opened and the policemen climbed aboard. Words were exchanged with the driver; the driver exchanged words with a Committee man; the Committee man exchanged words with the foolish teenager's dad; a clout on the back of the head and all was right with the world.

I marvelled at the pace and fair-mindedness of Tin Hat justice.

As we crossed the city and headed towards Wragby and Horncastle, the scenery began to flatten out, but it was no less spectacular. Mile after mile of winding tarmac, bound and guided by mile after mile of trees, fields, verges and hedgerows. Then, after Horncastle, the Lincolnshire Wolds carried us through Hagworthingham and Scremby on the A158; the undulating countryside finally levelling to a dead-straight horizon for the last few miles to the coast.

As we got closer to Skeggy, Butlin's holiday camp (Billy Butlin's first camp, and an immediate success when it opened in 1936) began to rise from that dead-straight horizon. That was the moment . . . a round of applause from the grown-ups and excited chatter from the kids, which eventually hit a high note as we pulled through the main gates of our caravan site. There was always a short stop at the reception as someone from the Tin Hat Committee went out to explain who we were and ask where we could find our caravans. Just enough time for the grown-ups to have a quick fag and the kids to mill about in front of the on-site supermarket, its winking, twinkling frontage loaded down with the bright oranges, greens, yellows, reds and blues of buckets, spades, lilos and assorted novelties.

Most shops also stocked a fine collection of ashtrays. Not just the usual plate or bowl-type affair that you found in the

Tin Hat. These were complex, mechanical marvels, using various sets of gears and hinges to dispose of a recently cremated John Player Gold Leaf Menthol . . . with the Cool Taste.

Ash would be tapped onto a shiny chrome plate or elegantly fluted opening, after which a button, knob or lever would whisk it away into a hidden chamber. The designs were as varied as they were ingenious: barrels, aeroplanes, animals, cars, buses, shoes. There was even one we had at home that was fashioned after a medieval monk. Tipping back his tonsured head would cause a large erect penis to spring from under his habit. Pull off the end of the penis, deposit ash, back on with his bellend, head forward, knob down and everybody's happy.

Remarkable things happened when we stayed in those caravans. Mam, Dad and me drank tea together. We listened to the radio together. Mam and Dad would sometimes walk arm in arm. We took photos. Instead of coming back from work, Dad would get up in the morning, same as everyone else. In fact, he got up before my mam and most of the neighbours. Every day at six thirty, Dad would quietly get dressed, then sit on the caravan steps, quickly smoking a quick fag. And coughing. After knocking on my bedroom door and waiting while I grabbed a bucket and spade, he'd roll his swimming trunks into a towel and we'd walk the hundred yards or so to the beach. Other men that Dad knew would arrive at the same time, perhaps half a dozen in total, all carrying a towel and trunks. Two of them would hold up towels while the others changed into their trunks; shirts, trousers and shoes neatly placed on their corresponding towels.

Pulling off my sandals, I found a decent castle-building/ comic-reading spot on the sand and watched as their compact, wiry bodies jogged towards the sea – which could take a while if the tide was out. Miners' bodies had a definite shape and style: legs short and bowed, hands like shovels, rangy tattooed arms, muscly without being muscle-bound. The other men even had skin like my dad's: nicked, grazed and marked with a thousand blue scars; almost tanned by the ground-in coal dust and the constant wear and tear. The older men also had a pronounced forward curve to the back and shoulders, a result of the cramped spaces their bodies had been wedged into on the coalface.

But here, in occasionally sunny Skeggy, none of that mattered. Aches, pains and dodgy knees were eased by the murky, not-very-warm waters of the North Sea. The men's laughter carried up and down the beach as they jostled each other like schoolboys, splashing, dunking and climbing on shoulders. Dad was a strong swimmer and would set off for some invisible marker he'd picked out on the water, his lazy front crawl carrying him quickly out to sea. At the allotted point, he would turn and race back towards his friends, one of whom had been lifted high above the heads of the others.

'*Weh-heyyy!*' they cried, launching the lifted bloke back into the foam.

Their fun and games continued as the morning sun finally began to grin at me from behind the long Lincolnshire clouds and a few more families made their way to the beach. Dad and a couple of the others appeared to be washing themselves in the sea, bobbing under the water before vigorously scrubbing their arms and legs with cupped hands. Head under water again, this time to give hair, scalp, face, ears and neck a lengthy pummelling. I had seen Hindu devotees on the telly, bathing and cleansing their souls in the river Ganges. Dad and his mates seemed to be imbued with that same spiritual delight, washing away sweat, dust and pain, while the water soothed their damaged skin.

Eventually, stomachs would rumble and the men would trudge back towards their shirts and towels, happy and glistening, ready for a day of markets, amusements and rock shops. There were a few more jokes and pulled legs as they dressed, but the mirth ground to a halt as they climbed the steps off the beach and headed back towards their caravans.

As if they were worried that wives or the wider population of Skeggy would see them enjoying themselves.

That wouldn't do, would it? Ostentatiously enjoying yourself on holiday?

The days took on a familiar pattern. Beach in the morning, breakfast in the caravan (fried egg, fried bread, fried sausage, fried mushrooms, fried Corn Flakes, fried milk, fried cup of tea, fried cutlery), then meet up with a few other families (Aunt Lal would be in our caravan) and walk across to the market. The same stalls were selling the same shoes, dresses, handbags, slacks, towels, suitcases, toys, sweets, records and T-shirts as the previous day, but we still stopped at every one of them. My mam would again wonder if she needed a couple of hand towels and Aunt Lal, again, told her to get them on Friday so they weren't cluttering up the caravan for the rest of the week.

After a bag of chips for lunch, we'd head for the town centre and more shops selling more of the stuff we'd seen on the market. The pace was leisurely in the extreme, with regular fag breaks and conversations with other men and women that we knew. At the top of Lumley Road (the main shopping street) was Grand Parade (the main promenade). As far as kids were concerned, this was . . . heaven. The lights, the colours, the noise, the smells. Amusements, toyshops, sweet shops and ice cream on one side of the road. A fairground, boating lake, donkeys, crazy golf and go-karts on the other. If we were dead and this really was heaven, we didn't mind.

Arriving at Grand Parade was also the cue for the men, women and children to split into groups. Children to the fairground or the amusements, women to some benches by the boating lake and men to a pub or, more often than not, a walk. Much as I loved the dodgems and the waltzer, I was also fascinated by Dad and his mates and their . . . walking. That was all they did. Walk and look at stuff with their shirts off or undone to the waist. I don't think they were showing off. 'Hey, ladies! Want a piece of this?' They were simply removing their shirts because it was sunny and they were hot. They were cool dudes without trying to be cool dudes. Just a bunch of men in their thirties, forties, fifties and sixties, languidly strolling down the pavement. Those long, lean arms swinging soft and low; perhaps a pre-rolled fag tucked behind the ear.

Although Dad and his band of Merry Miners gave off no hint of malice, I would often see groups of younger, louder men cross the road to avoid bumping into them. There was something solid and resolute about them and the other groups of shirtless miners we'd see during the afternoon. These were men who had spent their entire working lives in mortal danger. The snorting youngsters knew they would be safer if they looked for trouble elsewhere.

The best bit of the walk was when we got to the northern end of the seafront and turned left into some of the swankier roads like St Andrew's Drive and Sunningdale Drive. Dad and his mates were fascinated by the size and doll's-house neatness of these six- and seven-bedroom homes, but even more fascinated by the gardens. They, like most miners, were all keen gardeners and spent the rest of the afternoon peering through gates and over hedges, ogling rockeries, raised beds and dazzling roses.

Once, they spotted a bloke mowing his giant front lawn with a ride-on lawnmower. They found the other-worldly extravagance of the scene hilarious. A lawnmower with four wheels, reverse gear and a driver's seat! After much laughter and finger-pointing from our side of his hedge, the man rode over to us. He seemed to understand the situation immediately.

'*Duzz tha want a goo?*' He asked the question with a big smile and wide eyes.

We were there for . . . well, it was a long time. Long enough for every one of them to have a go on this futuristic gardening chariot. The owner, an ex-miner, relished the attention, pointing out all the buttons and gizmos, unashamed of the back-breaking hours this gleaming red beast had saved him.

Eventually, one of the men asked how much it cost. Not

brazenly. Miners were too polite for that. The subject of cost was casually introduced into the conversation and the owner revealed that he'd bought it with the insurance money after his wife died.

'It's wot shay would 'a wanted,' he added with an impish chuckle that gently shook his shoulders. Dad and the men joined in, gently shaking their shoulders, too.

Imagine if I'd stayed with the other kids and gone on the dodgems. I would have missed all of that.

Although Skegness was more than 80 miles from Dad's pit, he and several of his mates would usually be required to pack their St John uniform in the holiday suitcase. This meant there was likely to be brigade business while we were away: some marching, brigade gatherings involving beer, more marching and more beer. On certain Sunday mornings, men from all over the East Midlands would get dolled up in their uniforms and march around Skegness. The other kids and me would wait excitedly by the side of the road, hoping to catch a glimpse of a goose-stepping parent.

'Dad. Dad. Ay-upp, Dad. Can y'see me? Ha' y' bin upp t' t' church? Ah'm ovver 'ere.'

There were women in the brigade, too, and they would march in a separate column, just as proud and solemn-looking as the men. I suppose Mam could have joined, then she'd have had her own uniform to defile.

But she didn't.

There were occasional evenings out: a show on the Pier, maybe Tommy Cooper or Dick Emery. Occasional days out, too: a car boot sale, where I might find a still-in-the-box Action Man, an England football tracksuit or a decommissioned hand grenade. Sometimes, a couple of families would take the bus to Mablethorpe or Sutton on Sea. Dad and me would get off the bus at one of the pick-your-own fruit farms then, armed with a punnet of strawberries, meander down to the winding coast road that takes you north from Chapel St Leonards. Not the main A52; I'm talking about the smaller Anderby Road that leads into Roman Bank.

That road was not particularly long – 6 or 7 miles at most – but it was the most beautiful stretch of road I'd ever seen. A road made for adventure. A road like the ones in *Famous Five* books or *Pigeon Post* by Arthur Ransome. Banked by sandy bevels that would easily hide a casket of stolen treasure or a locked briefcase of top-secret government papers.

Looking north, the sea was on our right. On our left, caravan towns interrupted by ludicrously flat fields of cereal, grazing grass and rapeseed, each marked out by unruly hedges, occasional trees and tumbledown fences. A mishmash English landscape of odd shapes, sizes and colours, cobbled together over hundreds of years.

We passed signs to strange-sounding places like Wolla Bank Pits (submerged clay pits that were dug after the floods of 1953; the clay was used to repair the sea wall). Among the reed beds at Wolla Bank, Dad and me saw common spotted-orchids and wild celery (I would grab a couple of handfuls if nobody was looking), not to mention sedge warblers, herons and the elegantly eldritch great crested grebe. Further up the road at Huttoft Bank Pit, Dad pointed out great egrets, black-throated divers, water pipits, spoonbills, whimbrels, owls and the occasional osprey.

And the beaches. Anderby Creek, Moggs Eye and Chapel Six Marshes. Depending on the time of day, Dad and me might be the only people there, enjoying a couple of miles of rugged sandy coastline all to ourselves. In honour of our Dumbles trips, we didn't say much. We just stood. And watched. Pointing out a cormorant or sooty shearwater.

Apart from the odd bit of Byron and the wonderfully named Basil Bunting's 'What the Chairman Told Tom', I didn't care much for poetry. But I had read that Alfred, Lord Tennyson grew up near Skeggy and, as a young man, used to visit this same stretch of coast, shouting his poems into the same south-easterly that was now leaning its broad shoulders into me and Dad. I imagined his hoarse, wraithlike voice defying the elements . . . dancing across the sand, then skipping over the choppy waters of the North Sea.

Yet all things must die.
The stream will cease to flow,
The wind will cease to blow,
The clouds will cease to fleet,
The heart will cease to beat.

I only ever saw my dad cry once and it was on that coast in the summer of 1976. I know it was 1976 because I heard 'Jungle Rock' by Hank Mizell playing on a radio as Dad and me were walking past the small parade of shops at Anderby Creek. After climbing up the bank to the beach, Dad took off his cheap canvas trainers and sat down on the sand. For a minute or two, he looked out across the flat-as-a-pancake sea, then started rolling a fag. He rolled it nervously, as if he needed to give his hands something to do. I noticed his eyes were watering and, at first, assumed it was that ever-present broad-shouldered south-easterly blowing sand into his face. But there was something else in Dad's steady gaze. Something more than wind and sand. He was looking beyond the wind and the sand. Out to sea. Beyond the sea.

In between coughs, he shifted his gaze. Up the coast, down the coast, over his shoulder to the caravans, out beyond the sea. And up. To a capacious cerulean sky that was bigger than the coast, the sea, the caravans, me, Dad and anything else you could think of. So big, bright and overpowering that I hadn't even noticed it. A limitless playground for bands of scudding clouds and gangs of gulls. That's what Dad was watching. The sky, the clouds and especially the gulls. Hanging on unseen air currents, almost immobile against the sun. Or tipping and gliding, climbing and circling. Their lonely, singular cries slowly collapsing into harsh group laughter. Maybe laughing at my dad because he was crying. Maybe laughing at me because I was standing next to my dad in a tartan flat cap and woolly tank top on one of the hottest days of the year. Maybe laughing at all of us. Maybe just . . . laughing. Unable to contain the joy of hanging on unseen air currents. Tipping and gliding, climbing and circling.

They made Dad happy. And Dad was crying because he was happy. Much the same as I had been moved by the lapwings and rabbits when me and him were on the way to pick up his wages. The combination of North Sea, Lincolnshire skies, English countryside and laughing seagulls raised the hairs on the back of his rugged miner's neck and sent a shiver down his sturdy spine. And he was overcome by blissful, tearful contentment.

Behind me, I caught snippets of another sound, wafting on the Anderby breeze. It was my mam and Aunt Lal. Discussing the price of slacks and slippers on the market.

Time's up.

With a final glance at the sky, my dad fastened his cheap canvas trainers and walked off the sand. With a final glance at the sky, I followed him.

By that summer of 1976, my flying adventures – unlike the gulls – were almost over. I was eleven and had more or less worked out that those flying adventures were nothing more than dreams. Granted, those dreams had been powerful enough to leave me with the scent of high-altitude air still fresh in my nostrils and lucid enough to make my chest heave with the excitement of a perfect landing at Leicester Forest East Service Station. But they were still dreams and, like most eleven-year-olds, I had no time for such dreams. Too busy trying to steal a tip-top table-tennis bat from Vernon Hill. Too busy watching *Tiswas*. Too busy tucking into a bowl of Angel Delight.

But that night, I had a flying dream. About my dad. Dipping and soaring, just like me. Dressed in his favourite green jerkin, worn navy cords and cheap canvas trainers. A brand-new pair, though. Overflowing with sand and coal dust and stars and memories that trailed behind him like the endless vapour from a jet engine.

And he was laughing . . . like a seagull.

Exploration of Self

My years at Bagthorpe Primary School had opened some interesting new doors. I earned coloured braids for swimming. I was a regular in the football team. My love of birds became official when Mr Watson, one of the teachers, introduced me to the Young Ornithologists' Club. He also started an Astronomy Club, which led me to Patrick Moore's *Naked-Eye Astronomy*. I learned about light years, light curves, the 'movement' of the stars and planetary occultations. At night, my flying dreams had been replaced by actual journeys. After Mam had gone to bed, I would walk up to the M1 bridge at the top of Bentinck Lane and listen to the Doppler Effect: the satisfying 'meeeeowwwmmmm' of all sorts of vehicles as they sped past me. Or I would head for the top of my ex-flying hill. To the longer grass at the back; away from the brow and its views over the few lights of the estate. And I would lie down. Quietly. Looking. Quietly. Letting my eyes adjust.

On the last day at Bagthorpe, our teacher, Mr Hallet, wandered around the class pointing at people and telling them how well they were going to do in life. When he pointed to all of the other Clever Buggers, his face lit up. *'University!'* he cried. *'Definitely university!'*

When it came to me, I was expecting a similarly glorious fate.

Pointing. *'Maybe university!'*

I had no idea how, when or where I would go to university, but it would have been nice to have had a look at the dangling carrot before it was snatched away. I wasn't as friendly with Mr Hallet as I was with Mr Watson, but we . . . got on. I made him laugh with a fruit-themed sci-fi comic I drew called *Planet of the Grapes*. I impressed him with perfect spelling and maths test scores.

Admittedly, I was beginning to swear a lot – '*Fuck me backwards*' was the current favourite – and I had put Grovey in the hospital with my Judo chop. I had also recently told Mr Hallet off for smoking. Maybe it was that. As the head-master's office door swung open one morning, I'd spotted him inside, having a fag. Thinking I was being helpful, I told him about the book I was reading, *Common Sense About Smoking*, and quoted some of the terrifying lung cancer figures. His curled lip made me think I hadn't been as helpful as I'd hoped.

So what? Forget the lung cancer figures, just look at those maths and spelling scores!

Having said that, even if Mr Hallet had pointed to me and, while dancing a jig of delight, hollered, '*Oxbridge for you, my lad,*' it wouldn't have made much difference. In the final year at Bagthorpe, I had been asked if I wanted to sit the once mandatory eleven-plus exam. The eleven-plus was the Clever Bugger's Access All Areas pass. If I'd taken the exam and passed – like some of the other Clever Buggers – I might have got the chance to go to grammar school in Eastwood or Nottingham. But when I brought up the subject at home, the reaction was much as I'd expected. Dad said nothing and my mam set fire to the kitchen curtains, followed by: "*Ow y' goona get t' Nottin'ham every day? Ooh's goona pay forritt?*'

Being independently minded, I could have just done the exam and not worried about the consequences, but there was no way of getting around the practicalities of grammar school in Eastwood or Nottingham. Catching the bus to Alfreton or Sutton on my own was no problem, but I was not as familiar with Eastwood or Nottingham. There was one lad in Selston who did go to Eastwood and it obviously involved a hardcore approach to haircut and uniform. Attempting to spend five years at grammar school without the support and permission of at least one parent seemed . . . ill-advised. I might as well have said I was spending the next five years as a muntjac deer. Or a rubber dinghy. Or a hobbing foot.

But all was not lost. My new school, Matthew Holland Comprehensive School – whose playing fields hosted the Carnival – was also physically attached to the library. Going to Matthew Holland meant I had ready and instant access to the works of John Wyndham, Jack London and Pierre Boulle. Books about American history – the year was 1976, America's bicentennial year; I even learned the words to 'The Star-Spangled Banner' – Second World War biographies and bestsellers by John le Carré and Len Deighton.

Even here at this much bigger school, among 1,500 pupils, I continued to flaunt my Clever Bugger status. This was an era when the thirty-odd children in every classroom were divided into 'sets' and I was quickly placed in the top set for every subject. My mate, Dink, wasn't so lucky and found himself in something called 'Remedial'. It wasn't long before I started taking the piss out of Dink and anyone else who couldn't answer any questions that were thrown out to the class. I wasn't a teacher's pet shouter-out-of-answers. No, I sat at the back with the cool kids, where I would snort and guffaw. *'Everyboddy knows it wa' t' Crimean War.'*

In the first year at Matthew Holland, our class was given a long-term project called Exploration of Self. Part of the work was describing what my life was going to be like in the year 2000, when I was thirty-five. This was mine, in a nutshell: I got a job as a motorbike mechanic before being recruited by the SAS. I was trained in various martial arts and became the UK's top assassin, called in to do the work that no one else could do. There followed a detailed description of a mission I had recently completed in Russia, first torturing then killing a couple of communist agents. I included all the technical jargon: rabbit punch, roundhouse kick, aim for the groin, lock knife, fingernails, fag burns, severed artery and so on. The lad sitting next to me wanted to be a train driver.

My drawing had come on, too, and I included a beautiful pencil sketch of the Triumph TR6 used by Steve McQueen in *The Great Escape*. Apart from the tanks and aeroplanes of my Church of England Infant School years, I hadn't taken much notice of art, but at Matthew Holland I added it to my burgeoning portfolio of talents: maths, physics, languages, history, English and, thanks to the many years I'd spent cooking for myself, home economics.

The library helped. In the same way that the first explorers had made their way across oceans and continents, I was drawn to the different sections, shelves and alcoves. Drawn to Dylan Thomas, David Nobbs and *The Three Investigators* series. I read absolutely anything I could find on Uri Geller, fell in love with Carl Sagan and persuaded Rosemary, still my favourite librarian, to order in a copy of *Your Mysterious Powers of ESP* by Harold Sherman. At Alfreton Market, I found a book called *Knife Throwing as a Modern Sport* by Harry K. McEvoy and Charles V. Gruzanski, and bought a throwing knife off a bloke I used to see walking down in Bagthorpe Plantation. I have no idea if he was a true outdoorsman, but he certainly gave that impression. Army boots, full camouflage outfit, big backpack hung with picks and shovels. Sometimes, I'd spot him at the top end of the Plantation, putting up a rudimentary tent and cooking food on an open fire.

At twelve and thirteen, I was spending more and more of my time away from home. In the morning, I went straight from my paper round to school and persuaded the cleaners to let me into the library. From school, I would go for tea at Dink's or Bob D's. Often, I would even have a bath and do my homework at these 'other' houses. At weekends, I would stay at Collo's from Friday night to Sunday night. There wasn't any falling out or silly arguments with my mam and dad. Well, nothing out of the ordinary. No newly discovered anger about or interest in my adoption. No lingering resentment about them not being able to help me with schoolwork. No disappointment at not getting the latest Adidas sports bag or Wrangler cords. The older I got, the less I cared about those things. What happened was more of a gradual, glacial drift. Inch by inch, year after year. We became lodgers in the same house. Bumping into each other at odd hours and occasionally sharing breakfast.

The one thing that really rankled was the unpredictability of having my mam as a parent. As I was much taller than her, she was no longer a physical danger to me. What scared me and saddened me even more was the ever-present threat of 'the madness'. Setting fire to herself or the sofa or the house. Setting fire to the chip pan. Chucking away something I needed for school. Trying to move the dustbin on the back yard and falling over, cutting her leg badly. Cleaning the upstairs toilet and squirting Domestos on the floor, stepping in it and leaving bleach footprints around the house. Washing a friend's cat that I was looking after in Persil . . . thankfully, in the sink, not the washing machine. Smoking in bed and falling asleep.

I didn't even care about the big stuff any more . . . the eleven-plus or the Strap. (Ta-daa!) They'd been and gone. It was this constant drip-drip of low-key chaos. On their own, each incident was containable. But they were beginning to bleed into a dangerous whole, with bedlam knocking on our front door, day after day after day.

I'm sure that's why I became friends with lads like Dink, Bob D or Collo. Collo's house was 100 per cent bedlam-free. His mam's breakfasts were wonderfully bedlam-free, allowing us all to sit round the kitchen table, talking about coal tits and football. For dinner, she cooked plates of fish fingers, chips and peas that looked just like they did in the adverts on telly. I wished I lived at Collo's house.

One Friday night, I'd not been able to get to Collo's, so I turned up on Saturday morning. At quarter to seven. Collo's dad answered the door in a blue paisley-pattern dressing gown.

'*Danny-lad. Way love yuh, way really do. Burr it's a bit early. Yuh canna tonn upp at this time.*'

I could see Collo at the top of the stairs, looking down at us. An unsure look. Should he be annoyed, like his dad? Or happy to see me? And how was I supposed to react? I wanted to stomp off with a monk on and I could already hear a suitable sentence, itching to escape.

'*Ah'm gooin' homm.*'

If Collo came to knock on my door at five o'clock in the morning or even two o'clock, I would welcome him with wide open arms and a broad, beaming smile. By turning up early, I was putting myself in the same position and was obviously hoping for the same reception. I wanted Collo and his dad to understand just how important their house was to my life. A house full of nice wallpaper and nice carpets. The little table for the telephone, just the right size, sat by that bit of wall by the front door. They and their nice house could rely on me and my regular visits. Even in the depths of winter, when the snow and ice forced me to abandon my Raleigh Chopper and tramp through wild winds and snow-drifts, I would not let them down.

It was the wallpaper that did it. And the telephone table. It was Collo's mam's hair and her nice chips. His dad's ability to sit and watch telly with us or take us to Vernon Hill in Alfreton in his car. Still outside the front door, I started crying. Filled with shame and confusion. Collo's dad put his arms around me and took me into the kitchen. I hadn't expected that. More confusion. Collo's mam and his little sister came to see me. And more confusion. We all had break-fast together and did some birdwatching. We played with Collo's Subbuteo and watched the wrestling on telly. But when I went home on Sunday night, I knew things were never going to be the same again. And no matter how early or late I knocked on the door, Collo's nice parents were not going to let me move into their nice house.

Admittedly, I had other domestic options – I could stay with Dink or Bob D – but I was far more cautious now.

And caution turned into panic. Dink had three older brothers and, if I was eating dinner as well, we struggled to all get around the table. That now seemed significant. Everything seemed significant. I stopped having baths at other people's houses. Or doing my homework. Or changing channel on the telly so I could watch what I wanted to watch. In the end, I got tired of wondering if I was doing something wrong and simply acknowledged the endgame.

Initially, I struggled. And I wondered why I struggled. All I had to do was live at home. I had lived with my mam and dad for my whole life. They weren't my real mam and dad, but that was hardly their fault. Without them, I would have ended up in the orphanage thingy, which – having read about orphanage thingies – I now understood to be nothing like a school.

If I had transgressed at Collo's front door, it was my own silly mistake. If I had overstepped the mark at Dink's or Bob D's, it was caused by my own inadequacies. And the best way to remedy those mistakes and inadequacies was by hard work, experiment and everyday trial. To challenge myself and to always be victorious. So I braved the elements and – like the man in the Plantation – camped out under the stars, at the foot of the Canyon on my flying hill. Or at the top of the field where the horses lived. I adopted a hardcore mental and physical fitness routine that included buying 99s from the ice cream man and throwing them away (to prove I was in control of my primal instincts), jogging barefoot after dark (conquering pain and cold), reading at least three books a week (brain power), staying completely still for anything up to an hour while concealed in a gorse bush or watching telly (hiding from an enemy), remembering ten things about every house on our street (memory skills), making a meal with only one arm (in case I lost an arm during combat), holding my breath (useful if I was trapped in an underwater cave or caught in a gas attack) and learning ventriloquism (haven't got a fucking clue).

I took home bits and pieces from the physics lab, taught myself the basics of electronics and stole methylated spirits from the garage at Wilde's Corner so I could make Molotov cocktails. Nokky Hancock's judo classes had finished, but I signed up for karate, which took place in the sports hall next to Matthew Holland school. As described in E. J. Harrison's *The Manual of Karate*, I wrapped rope around the concrete washing-line post by our coal shed, creating a makiwara or 'striking post' to harden my fists and feet. By the time I left Matthew Holland Comprehensive, I wouldn't just be a Clever Bugger. I would be the Cleverest Bugger in Selston.

And possibly the most dangerous.

I continued to wrestle with my mistakes and inadequacies. Punching and kicking them. Bamboozling them with information and technology, logic and reason. How could I make up for these mistakes and inadequacies? Maybe I couldn't . . . but The Mystery Man could.

Obviously, The Mystery Man wouldn't be allowed to wander around Selston willy-nilly, righting wrongs and dishing out summary justice whenever he felt like it. He needed officialdom and the law on his side. My letter to the main police station at Kirkby-in-Ashfield briefly described my plan. Stuff was going wrong in Selston and it needed sorting out. I gave full details of my training regime, my technological nous and the books I'd studied. I offered to help our overstretched police force, keeping an eye on things for them. Would it be possible to have a special phone line installed that would connect me directly with the chief of police? And could they furnish me with some sort of identity card that would let the authorities know I was working in tandem with our boys in blue and had special dispensation to dish out whatever punishments I deemed necessary?

A week later, there was a knock on the front door of our house. I'd seen the policeman get off his motorcycle and walk down our steps, so I made sure I was first to the door, fully expecting that he was there to hand over my ID card and ask me where I wanted the special phone. In my bedroom, I thought. I'd also decided to ask for a police radio so I could keep up to speed with burglaries, car crashes, murders and so on.

Unfortunately, my mam had heard the knock and came bustling into the passage.

"Oo is it? 'Oo is it?'

He was nice enough. Tall and softly spoken. Talking me and my mam through my letter. Sadly, no ID card. No phone. Not even a police radio. He asked if I wanted to join the police when I left school. I wasn't sure I did, but I said yes.

My mam had glazed over by this point, but she managed a couple of garbled complaints.

'Wot's ay bin upp tew? Ay's bin in trubble before, y' know!'

Far from being in trouble, he told my mam that I wanted to help the police.

'Y' wot? Ay wants t' join t' police? Ay hanna left school yit!'

Sitting on the sofa, he earnestly explained that crime fighting was best left to the police and he'd hate to see me getting myself into bother. I agreed and he offered to send me some information booklets about joining the police. At the door, we shook hands. Gripping and pumping hard, he put his other hand on my shoulder.

'Lukk after y' senn, lad. Dunna doo owt daft.'

I promised I wouldn't. But I lied.

The Mystery Man made several appearances that year. At his most successful, he swiftly dealt with the bloke on Gill Street who was letting his dog shit outside our gate. It wasn't always directly outside our gate, sometimes it was two or three doors up or down. At first, I scooped the dog shit into the hedge bottom or into the road, but it kept happening. Then, one evening, I found myself walking behind a man and a dog, just as the dog did his business near our hedge. Instead of turning into our gate, I continued walking . . .

slowly, keeping some distance between us. He turned left at the end of the jitty at the top of our road, down the hill and left into Gill Street. I waited, made my way to the corner and watched him go into a house. I couldn't be sure which one, so I casually walked past the place where he'd disappeared. A front room light was on and I saw both him and the dog.

That night, around eleven, I dressed for danger. And took my mam's rubber gloves from underneath the sink. I'd moved Gill Street Man's dog shit under our hedge for safekeeping and used the gloves to pick it up. Keeping to the shadows, I half jogged to the corner of Gill Street. All quiet. The street lights were poor but they convinced me to approach from the back garden. Through a gate, over a hedge, another hedge, a fence, another hedge or two. That was the house. Dark. A smear of dog shit on the kitchen window. A smear on the back door handle. On the coal shed handle. On the front door handle. And the rest scraped on the wall. Then back down the garden, over hedges and fences, up the hill, back in the house. I tried washing the dog shit off the gloves, but they still smelled so I dug into the dustbin and buried them.

I don't know where the dog did his business after that, but it wasn't on our road.

As The Mystery Man, I also decided that I would offer the hand of friendship to Mr Keith Kent and His Invisible Wife. It was easy. Wait until I'd seen the lights go out in his front bedroom; slip on the overcoat, hat and scarf. Across two front gardens, through a hedge, over a hedge, up to the front door. In black felt-tip capitals, just under his letter box:

I TRUST THIS FINDS YOU AND YOUR WIFE WELL.
YOUR SECRET IS SAFE WITH ME.
A FRIEND.

From my window, I looked across at the front door the following morning, but the words had already been rubbed off. Smart move, Mr Keith Kent. Let's keep this one between me and thee, eh!

Unfortunately, for every success there was . . . disappointment. Despite rigorous testing, the Molotov cocktail I made to blow up Mr Rowland's shed didn't work – impressive to look at, but burnt itself out far too quickly to set fire to anything. And I cut my arm when I got caught on a barbed wire fence, climbing into Wayne Gilbert's back garden to check if he'd nicked the front light off my Raleigh Chopper. A nasty cut that Dad had to re-dress the following morning.

Maybe at thirteen, I was getting too old to be a superhero. And maybe I was still too young to be the Most Dangerous Bugger In Selston.

That didn't stop me being the Cleverest Bugger, though. And with The Mystery Man retired, I focussed all my efforts on the acquisition of knowledge. Books, magazines, Open University programmes on telly. And it sort of worked. There were times when I sat in a physics or maths lesson and truly felt at home. English lessons, too. The first time I read Keith Waterhouse's *Billy Liar*, I sensed my own literary imagination soaring towards stories of my own. And table tennis; after all that time studying *Know The Game: Table Tennis, Published In Collaboration With The English Table Tennis Association*, I had a ferocious and lightning-quick backhand. Helped by the tip-top bat I eventually stole from Vernon Hill.

Granted, there were a couple of clever lads and lasses at Matthew Holland who could match/outdistance me in single subjects. Andrew Brown, for instance, was a physics genius, but he had no idea who sang the theme tune to the 1977 Bond movie, *The Spy Who Loved Me*. Lorraine Barton wrote wonderful stories, but she had no idea of the difference between a two- and four-stroke engine. As an all-round Clever Bugger, I was pretty much in a league of my own.

In my final year at Matthew Holland, I was picked for a team in the annual school quiz. I had to wear my uniform, but I was now old enough to get away with a blue crew-neck jumper, no tie, black cord trousers and monkey boots. It was a special night. I answered questions that stumped the teachers on our team. Sport, songs from the Second World War, telly shows, American history, the laws of physics, literature, fashion, food. I answered questions that weren't even directed at me, nonchalantly buzzing in with the full name of TNT (trinitrotoluene, $C_6H_2(NO_2)_3CH_3$) or pi to ten decimal places (3.1415926535).

Neither my mam nor my dad had ever been to a parents' evening, so I never bothered mentioning the quiz night. Had they been there, I'm not sure I would have been able to hold my nerve. Too worried about my mam blurting out something about me wetting the bed. Or blurting out something about me being dumped in an orphanage. Or blurting out some awful double entendre if one of the questions included a word like 'handle', 'hard' or 'erect'.

She did find out about the quiz, though, because my teacher, Mr Hall, went to see her a few days after my triumphant performance and told her about something called the sixth form. My dad was still on the night shift, so he was in bed, but I don't think it would have made much difference even if he'd been there. All he would have heard was Mr Hall valiantly trying to discuss which A levels I ought to be thinking about – after watching Carl Sagan's groundbreaking series, *Cosmos*, I quite fancied being an astrophysicist – while my mam trundled awkwardly around the kitchen like a dalek with faulty electrics.

'*It's abaht time ay started earnin' summ money,*' she told Mr Hall. '*Way canna keep payin' forrim.*'

Foolishly, Mr Hall decided to put up a fight. He talked about the help that was available for families like ours. He talked about grants and scholarships. He offered to speak to the right department at the council and sort out all the paperwork.

He was wasting his time.

Mr Hall even tried talking about vague, wishy-washy concepts like ability, achievement and getting on in life.

170

He was still wasting his time.

At one point, Mr Hall actually said, '*Mrs Scott, I'm begging you.*'

Now he really was wasting his time.

My mam's trundling became ever more agitated. '*Exterminate,*' she shouted in a funny, mechanised voice. '*EXTERMINATE!*'

I Didn't Dwell on These Questions at the Time, but Many Years Later . . .

Had life in Selston always been like that? Had people like my mam and dad always been afraid of wishy-washy concepts like ability, achievement and getting on in life?

During the difficult years before and after the First World War, I'd always imagined that knowledge was welcomed into the Selston miners' lives like a roaring fire on the coldest day of winter. Working-class people were finally given the chance to learn; and they wanted to learn. Mining communities began setting up their own libraries; five schools opened in the Selston area between 1840 and 1880. Newly literate Nottinghamshire mining men began making their mark on the national stage . . . flat-vowelled, straight-talking MPs, union leaders, peace campaigners, magistrates, headmasters and school governors. Men like William Carter, a former miner who was elected Labour MP for Mansfield in 1918; George Spencer, another former miner, elected Labour MP for Broxtowe in the same year. Men like Matthew Holland.

Most people in Selston knew about Matthew Holland. His family still lived in the village and his name sat proudly next to the entrance to my school and much was made about how he'd campaigned to build a decent education for the lads and lasses of Selston, Jacksdale, Underwood, Bagthorpe and Brinsley. A shining example of what simple lads and lasses from out-of-the-way mining villages could achieve.

The library had several copies of 'Matt': From Mines to Minds by David Wheatley. The one I picked up had a simple red cover and was signed by the author, dated 1967. According to this biography, Alderman Matthew Holland, CBE, JP, was born in 1872 and went to Selston Church of England Infant School, like me. At eleven, he started work at the pit, eventually joining his father at Top Mexborough, one of the mines that lay buried underneath the estate I lived on. To collect his first week's pay of 2/9d, he had to walk from Selston to Eastwood and back, a round trip of ten miles. At just twenty-one, he was attached to the executive council of the local miners' union, often speaking out against corruption and victimisation in the coal industry. His reputation as an agitator eventually cost him his job, but he continued to fight the miners' corner until his death in 1957.

Matthew Holland's support for the miners was matched by a passion for education and he was instrumental in the transformation of the schooling available for Nottinghamshire's underprivileged children. As well as his campaign to build our school, he had been a pivotal figure in the Selston branch of the Workers' Educational Association – a national organisation, set up to bring education to 'the people'.

I suspect that reading Matthew Holland's story was my attempt to stock up on ammo . . . to be able to argue my case. Without much in the line of money or privilege, he'd gone from pit work at eleven years old to chairman of the Nottinghamshire County Education Committee. Matt and me were cut from the same cloth; we even went to the same infants' school. He worked at the pit that now lay beneath our estate. If I'd taken my eleven-plus or stayed on for the sixth form, wouldn't I be following in Alderman Holland's hallowed footsteps? Continuing the Selston miners' long-established tradition of self-improvement. The plucky collier's lad who battled against the odds. Shit shoes and a crap haircut but blessed with a mind as sharp as one of those knives in that book by Henry K. McEvoy and Charles V. Gruzanski.

Our other local hero, D. H. Lawrence, came from a mining family, too, and he won a scholarship to a posh school. Somebody stumped up the money that allowed him to learn!

He even mentions Selston on the opening page of his story, 'Odour of Chrysanthemums'; that's how close he was to me and Mam and Dad and our house. And our life. Wasn't me taking my eleven-plus or staying on for the sixth form or being a Clever Bugger something to be encouraged? Something to be celebrated?

There was no point bringing up D. H. Lawrence with my mam, but I thought there might be a slim chance that Matthew Holland could change her mind. I ought to have known better.

When I mentioned Matt Holland, she seemed puzzled. *'Wot yuh talkin' abaht 'im for? Ay's dedd, in't 'e? Ay's non goona 'elp yuh. Ay canna gi' yuh a job.'*

After sixteen years, I really, really ought to have known better.

And, in a way, she was probably right. The sad truth is that, even while Matthew Holland was alive and introducing the first WEA classes in 1913, there were some miners who consistently blocked their own, and their children's, chances of advancement and learning. Thoughts of Lawrence had also reminded me of *Sons and Lovers* – published in 1913, the same year as those first WEA classes in Selston – and Walter Morel's reaction to the news that his wife has secured Paul's brother, William, a job in the Co-op office.

> *'What dost want ter ma'e a stool-harsed Jack on 'im for?'* said Morel. *'All he'll do is to wear his britches behind out an' earn nowt . . . It* [working at the pit] *wor good enough for me, but it's non good enough for 'im.'*

Although he was only a character in a book, Walter was based on Lawrence's own father. Indeed, many of Selston's real-life Walter Morels seemed to share his unease. Following the 1926 General Strike, the miners began distancing themselves from Matthew Holland and the aspirational hopes of the WEA. With their lives increasingly governed by hunger and fear of unemployment, they had little time or need for lectures on philosophy or constitutional law. Such luxuries were best left to them that had money and time on their hands. Best left to local do-gooders like Alderman Matthew Holland, CBE, JP.

Even after the Second World War and the nationalisation of the coal industry – accompanied by the sweeping changes of the 1944 Education Act – there were large sections of the mining community that remained suspicious of anyone who talked about ambition, office jobs or passing exams.

Mr Hall, the teacher who'd tried to change my mam's mind, put me in touch with one of his former pupils, Kev Patterson, a former miner at Bentinck who'd gone on to be a draughtsman at an engineering firm in Mansfield. Kev told me about his own father's reluctance to apply for promotion at the pit where he worked. Kev's dad was a face worker; his friends and neighbours were face workers. Understandably, he thought – like generations of miners before him – that applying for promotion would be seen as a betrayal of the people he worked and lived with. As if he was telling them that their world was no longer good enough. Even worse, applying for promotion might even get him thrown off the darts team!

On the day Kev left school in the early '70s, his dad was waiting for him at the school gates, having already sorted out a job for his son at Bentinck Pit. After listening to Kev, I realised that it wasn't the least bit surprising that Mam and Dad had turned down Mr Hall's kind offer of two more years in education.

At sixteen, disappointment is a difficult emotion to even corner, never mind conquer. Especially this kind of disap-pointment. It wasn't like losing a football match or missing last orders at the chippy. Their immediacy made these disap-pointments easier to spot, easier to immerse yourself in . . . easier to put behind you. But being prevented from learning some more stuff that may or not serve you in good stead in ten or fifteen years' time has a hazy, half-formed feel to it. Yes, I might end up following in James Christy's footsteps, harnessing my interest in astronomy and getting a job at NASA, discovering a second moon orbiting Pluto, but I might not. I might end up harnessing my mechanical skills and working for a Suzuki motorcycle Grand Prix team powered to victory by an engine that I designed, but I might not.

Even if I did end up disappointed, it wasn't going to happen for a long time. And a long time is a VERY long time when you're sixteen. Being told that, in ten years' time, you might not get a job at NASA or Suzuki wasn't exactly life-changing. A lot can happen in ten years. Maybe I wouldn't even need all those qualifications to work at NASA or Suzuki. Even without the sixth form, I was still me . . . I was still, thanks to the full name of TNT and pi to ten decimal places, the Cleverest Bugger in Selston.

I soon shook off the remnants of any hazy, half-formed disappointment. Almost as quickly as my mam and dad shook off the crazy, half-arsed idea that their son might work for NASA or Suzuki. Two more years when they'd have to provide me with free board and lodging? Two more years of books and those pesky teachers? Two more years that might get Dad – as he was never a keen darts player – thrown off the Tin Hat Committee?

My dad's wages had just about kept us solvent; now it was my turn to start chipping in. The gravy train had finally ground to a halt.

'*Tha mun think thee sen lukky,*' Mam reminded me. '*Ah were in service at fourteen. An' y' dad were wokkin' at t' pipe yard.*'

To say my mam had a blinkered view of the world and what it had to offer is pointless. She was blind . . . literally and figuratively. Her imagination was almost as useless as her eyes. And even the little bit of imagination that was still on tap had gone into self-preservation mode. No point imagining that you could use a guide dog to give you greater independence because something would go wrong and people would laugh at you. No point imagining that your husband might become a pit deputy because something would go wrong and people would laugh at him and you. No point imagining your son might work for NASA or Suzuki because . . . well, does it really matter?

One evening, after my dad started doing the pools, we were sitting around the kitchen table and I wondered aloud what we'd do if he won the £75,000 jackpot. I said I would buy an off-road motorbike. Dad fancied a rotavator to help him in the garden. Mam thought for a few seconds and grandly announced, *'Ah've gorra sayin'.'*

Dad and me looked at each other, eyebrows raised in anticipation of these pearls of wisdom.

Mam flibbed her fag and cleared her throat. *'There's no point wishin' fo' owt.'*

Obviously, Dad and me kept quiet, waiting for her to finish off with the wise words that would undoubtedly prove Mam was related to Confucious and we were going to win the pools.

We waited.

Waited.

A bit longer.

Then, I understood. That was it. DON'T WISH. For people like us, they were the only wise words we'd ever need.

There was a brief moment when I thought Dad might stagger to his feet and say something about NASA and Suzuki. In his own way, he had encouraged me. It hadn't happened many times, but it did happen. When he knew I was interested in dinosaurs, he brought back a trilobite fossil that had been found at the pit. When he knew I was interested in birds, he got me the *I-Spy Birds* book. When I started collecting the football cards that came with Typhoo packets of tea, he got some of his mates to save them for him and handed over a small bagful that included the much-sought-after Derby County striker, John O'Hare. When I took up martial arts, he came home with a pair of boxing gloves.

But NASA and Suzuki were a bit different to Typhoo tea. Dad said nothing and I couldn't blame him, really. Here was a man who'd lived in Selston his whole life, had never been abroad and only left the East Midlands for his honeymoon in Blackpool, a day trip to Bridlington and the occasional far-flung St John Ambulance get-together in Rhyl.

It was easy to see why a modest, uncomplicated man like my dad, living in a tightly knit community like Selston, wasn't keen on me getting ideas above my station. It was easy to see why Kev Patterson's dad was so anxious about his place on the darts team. And why my mam was scared shitless about me staying on for the sixth form. A promotion, a suit or a place at university might cause all sorts of mayhem. This group of people might stop speaking to you. The landlord of your favourite pub might decide that you were no longer welcome. Your younger siblings and relatives might suddenly find themselves being bullied by the older children of the people you'd upset with your exams and your highfalutin ideas.

In 1970s and early '80s Selston, such social mores, as petty and ridiculous as they may have seemed, were crucial to everyday life. They existed as the powerful stitching that was just managing to hold together the fraying fabric of Britain's mining communities. They maintained order. They kept everyone's garden neat and tidy, allowed them to leave their back door unlocked and gave the world a soothing familiarity that made it so much easier to deal with.

Coal had given Selston a very specific, hard-wearing identity but, in return, it demanded and, to a great extent deserved, our loyalty. A brutish but wholehearted and everlasting loyalty. Coal had built both the houses I'd lived in, the three schools that had educated me, the shops I shopped in, and it weighed heavy on almost every branch of almost every family tree of almost everybody I knew. It shaped the language I spoke, the clothes I wore, the friends I made and the food I ate.

Working at the pit was much more than a 'job'. Like the church or medicine, it was a calling. It was a badge of honour. A set of beliefs. A ritual. Selston was packed with families that could trace their mining roots all the way back to a time before the world's first iron bridge in Shropshire or the world's first mechanised cotton mill in nearby Birmingham. Families that had endured hardship upon hardship and felt explosions rip through the very earth beneath their feet. They had mourned one death every eight hours and lined their streets with sadness. They were worried sick when men weren't back from their shift at the usual time. They had watched health and happiness drain from the faces of their own children.

Admittedly, other industries had similar ties to history and community: the shipyards, railways, steelworks, fishing and farming. Life-threatening industries that nevertheless drew down a long line of proud sons and fathers. And those industries all had their own worthy tales to tell. But fishing was not mining. And the shipyards were not mining. They were not bathed in the same damp, dusty, suffocating darkness.

Yet the miners carried on. And their children carried on. Generation after generation after generation. Right up to me and the many lads who were due to leave school in 1981. The intoxicating scent of sacrifice . . . those lads' sacrifice, my family's sacrifice. The rolling momentum of 'all that history' nudging and knocking us towards Bentinck, Huthwaite, Moorgreen or Pye Hill. Colleagues, neighbours and friends, all joined together by coal . . . the Undisputed King of Selston.

Mining wasn't just *a* way of life; it was the *only* way of life. It was who I was and who I was always meant to be.

The Circle of Coal

As I turned sixteen a few weeks before I was due to leave Matthew Holland Comprehensive, I was able to get a provisional driving licence. At weekends, I now worked the late-afternoon/early evening shift at the garage that was opposite Wilde's Corner. I served customers, filled their petrol tanks, took their money, sold automotive sundries and read books. The job allowed me to buy a Suzuki AP50; it cost me 20 quid from a lad in Pinxton and only just worked. Using our back yard and the strip of concrete wasteland at the side of the garage, I spent several weeks stripping and rebuilding the engine, getting rid of the rust and polishing the bits that needed to be shiny.

Then, one fine Saturday morning, I finally wheeled it up our front steps, pressed my foot down on the kick-starter and decided to run away. Well, ride away. But where to? Nottingham? Down the A38 to Derby? Through Alfreton and Clay Cross to Sheffield? What about my end-of-year exams? And my new job at the garage? I couldn't just ride away from them as well.

Playing it safe, I rode to South Wingfield, which is just the other side of Alfreton, on the way to Matlock Bath. Within a few minutes of leaving the estate, I was buzzing past the Pit Houses, the Clay Heaps and Dad's old pit, down Jubilee Hill with its sprawling meadow, past the shop where I bought milk and fags for my mam, through Somercotes, Alfreton and the swanky houses on Wingfield Road. At the

Peacock Inn, I turned right, then almost immediately left onto Dale Hill.

That is the start of the Peak District and Dale Hill is even more picturesque than its name suggests. It goes down and up and down and curves left and right and shoots off over here and there and does a bit of levelling out as it crosses the River Amber. I turned left onto Birches Lane, over Birches Brook, up through South Wingfield, past the Zion Methodist Church, across Boggy Brook and there it was, over on the left. Wingfield Manor.

I liked the Manor. It was another of the area's 'bits of history'. A proper bit. Commissioned in 1441 by Ralph Cromwell, it took so long to build that when he died fifteen years later, it still wasn't finished. From 1569, it was used as a prison for Mary, Queen of Scots. In 1596, the Manor took delivery of England's first flushing lavvy. Sadly, the building was severely damaged during the Civil War and had been deserted since the late eighteenth century.

It was guarded by a fence and gate, but it was never locked, so I was able to ride my gleaming Suzuki motorbike down the dirt track and right into the Manor's former courtyard. I could get off my gleaming Suzuki motorbike and wander into every corner of those ruins, including the set of stone steps that took me to the top of the fifteenth-century High Tower. Traffic on the main road was light, with five- or six-minute slices of time when I neither saw nor heard a car or even another human being. If I positioned my head at just the right angle and closed one eye, I could block out every scrap of evidence that tied me or the landscape to 1981.

I remembered my long-gone boyhood flights and wondered what this timeless landscape would look like from above. But how could I get up there from Wingfield Manor? I rested the palms of both hands on the ancient stonework, hoping to feel a spark of ancient sorcery.

Nothing.

I rested my forehead against it.

Still nothing.

I kissed it.

Nope.

As I had usually taken flight at night, it seemed likely that I would need to be in bed or at least in my pyjamas? But I no longer wore pyjamas. So I took off my jacket and stripped down to a pair of baggy Y-fronts. Was that a hint of something magical? I closed my eyes, spread my arms and chanted a few suitably age-old phrases: *'Fare thee well. Gadzooks, I am not restful here in this place.'*

It was not quite summer, but an unusually warm breeze suddenly whirled around the High Tower. A strong breeze that felt like it was coming from every direction at once. Strong enough to carry me and my gleaming Suzuki motorbike into a new world. A new world that hadn't gathered up my carefree past and used it to map out and sign off my future before sending it to the printers by registered post. A new world where there was room for manoeuvre.

I felt lifted. I felt dizzy. I felt hopeful. No, I felt sure. Sure that when I opened my eyes, I would see the new world I longed for. A world that appeared plain and unhurried. A world where the luxuriance of the grass and soft shape of the land told me there was nothing to fear. A world where England's first flushing lavvy was in perfect working order. Yes, the mining industry was still dotted willy-nilly across the East Midlands but I somehow thought that this new world – this new, old world – would offer me a few more options. The past had given the nod to mining lads like Keir Hardie, Richard Trevithick, George Stephenson, even Nye Bevan, architect of the NHS. Surely, I would be in with a shout.

I slowly opened one eye – just the one. No sign of 1981. In a short prayer, I thanked the Irish philosopher, George Berkeley (his influential 1710 work, *A Treatise Concerning the Principles of Human Knowledge*, had been a regular comfort in my last year at Matthew Holland). Berkeley was the man responsible for the theory of immaterialism and had raised the question of whether the material world exists if no one is there to perceive it.

If a tree falls in a forest and no one is around to hear it, does the tree make a sound?

According to immaterialist principles, the fact that I could no longer see 1981 meant that 1981 had ceased to exist. Here was my chance to join Keir, Richard, George and Nye.

But the longer I looked, the more I doubted. Even though my view from the High Tower was joyously unshackled from the 1981 calendar, 1981 and all the other years in between seemed to be sneaking back over the horizon. At first, it was nothing more than a faint but familiar sound echoing across the Peak District hillside. A simple and insistent two-note cry of pain. Bee-Daw-Bee-Daw-Bee-Daw! Followed by the metallic whine of an overworked car engine and a clumsy crunch through the gearbox. Down the track it came, blue lights flashing. Demented wheels churning away at the innocent earth. A fine, honourable earth. A kind and wise earth that had once carried the bejewelled boots of kings and queens.

Although I couldn't see 1981, it could see me. Quite clearly. From a house opposite the Old Yew Tree pub on Manor Road. A house with a phone. Luckily, the fresh-faced constable agreed that, having just restored my gleaming Suzuki motorbike, I was unlikely to be considering suicide. And as I was in my Y-fronts, he could probably cross off vandalism and Satanic Ritual. We even shared a joke as he waved me down the road.

Which road? Did it matter? It would be a road that led to Selston. And, in 1981, all roads in Selston led to . . . the pit.

Admittedly, not every road in Selston ended at a pit, but every road in or out of the village would, without fail, take you past the site of a pit. Like Avebury and its prehistoric stone circle, Selston was once surrounded by winding gear, spoil heaps, engine sheds, workshops, administrative buildings, canteens and Clay Heaps. And like that circle of sandstone blocks in Wiltshire, the pits and their pit gubbins cast a bewildering spell over the entire village.

Once inside The Circle of Coal, you would hear the same sounds and words constantly whooping and whispering through the air. The names of the collieries: Bentinck, Pye Hill, Portland, Mexborough, Moorgreen, Bull & Butcher. The eerie whirr of the winding gear. The laboured throb of lorry engines. The muffled, murmured banter of long-dead men, women and children. The steady squeal of coal-cart wheels, thirty or forty of them shuffling and shunting behind a dirty blue diesel locomotive.

I heard them all, floating in and out of my bedroom window on a fairly early morning in August. As it was a fairly quiet fairly early morning, they were mixed in with birdsong and the faint throb and thrum of traffic on the M1, half a mile away. The summer was almost over and I'd spent most of it doing little more than tinkering with the gleaming Suzuki, working shifts at the petrol station, and growing my hair and as much of a beard/moustache as I could muster. I'd also bought a wafer-thin leather(ish) jacket and some light beige cowboy boots from Sutton Market in a vague attempt to look like Dennis Hopper in *Easy Rider*.

Having accepted that the sixth form wasn't going to happen, my ties with Matthew Holland were technically severed – apart from today's final journey to collect my exam results. Some lads and girls made a fuss of results day, going up there in excited threes and fours. The girls to share the joy/sadness, the boys to simply take the piss no matter how well/badly they'd done. There were other lads and girls who obviously had no interest in their results because they'd already secured positions on the family farm, the local textile factory or . . . the pit.

I was far less nervous than I thought I would be. Maybe it was because I was convinced I was a Clever Bugger and the results were already in the bag, or maybe it was because the ups and downs of the last few months had convinced me that my results didn't matter. O levels, A levels, a degree and doctorate from Harvard; it really did not fucking matter. Not a jot. Or an iota. Or even a jot of an iota.

I buzzed up Nottingham Road in full *Easy Rider* regalia, finished off with an AC/DC T-shirt and a pair of sunglasses somebody had left at the garage. I parked outside the library, took off my helmet – not the sunglasses – unzipped my wafer-thin leather(ish) jacket and headed for the entrance at the right of the library. The school was quieter than usual, a strange silence that amplified the clackety sound of my cowboy boots slowly counting down the steps along the corridor's parquet flooring.

My O level results were as good as I expected, though I did have to take my sunglasses off to read them. I was, without doubt, a Very Clever Bugger. Coupled with the regular Suzuki tinkering and ever-present smell of engine oil, the glorious grades in maths, physics and engineering gave me an idea. Apprenticeships were all the rage in 1981 and the National Coal Board (NCB) was well known for providing one of the most respected apprenticeships in the country. I tugged at my wispy beard and wondered. Granted, it wasn't NASA or Suzuki, but it would mean machines and books and knowing your way around a workshop. All the stuff I enjoyed. One bloke who lived up on Portland Road had started out as a fitter at the pit and now worked at the main Ford garage in Derby. Not NASA. Not Suzuki. But . . . not bad.

I mentioned this to my dad and, as usual, he said nothing. But he did stop rolling his fag. And he slowly nodded his head. Then nodded some more and took a swig of tea. In lieu of Typhoo tea football cards, I took this as a good sign. Dad's peculiar version of encouragement.

The mining wheels were in motion.

Dad talked to his mate who talked to his mate who agreed to have a talk with me about working at Bentinck. In preparation, I went to the library and read up on current coal-mine technology: from the narrow-gauge trains that carried the men to and from the coalface underground to the frighteningly fierce fully automated trepanners. A few days later, a letter arrived inviting me in for an informal chat. Reading between the lines, I felt this 'chat' wasn't so much an interview as a final inspection. The fact that I was my dad's son was all they needed to know.

I could have walked to Bentinck, but I liked bragging about the work I'd put into my gleaming Suzuki. It started first time and, as I rode towards Bentinck, I began to feel . . . excited. Relieved. Happy. Why hadn't I thought of this before? An apprenticeship. A fitter at the NCB. Dad wouldn't have to worry about me turning into a 'stool-harsed Jack' and wouldn't even get thrown off the Committee. Mam wouldn't be able to complain that she and Dad had to keep 'paying for me'. I would hand over a portion of my wages every week and she could spend it on fags and fire extinguishers. Maybe she, Dad and King Coal had planned this circuitous journey all along, gently guiding me towards destiny.

By the time I hit Bentinck Lane, my feet were well and truly under the table. I had reckoned that being a Very Clever Bugger would allow me to quickly and easily rise through the ranks of Nottinghamshire's mining engineers. I would be sent up and down the country, giving talks to other mining engineers about my revolutionary ideas for underground transportation. Within a couple of years, I would buy a house on the leafy lane that takes you past The Shepherds in Bagthorpe, the house with the stretch of grass and the line of trees that stood next to the brook. Plenty of room for cars and motorbikes . . . a souped-up signal-orange RS2000 Escort in the garage, alongside a stock GS 650G Katana, like the one that bloke on Annesley Lane had. I'd filled his tank up a couple of times at the garage and he'd even let me have a sit on it.

By the time I was sweeping round the right-hander, halfway down Bentinck Lane, I had started designing and decorating the house. Music room, martial arts training room, gym, no telly, extend the garage at the back, with space for one of those motorbike lifts I'd seen in that workshop in Mansfield. I'd do most of the work myself, of course.

By the time I actually parked up my Suzuki, I'd already

been given an MBE for services to industry. To celebrate, I invited Dad – maybe Mam as well – over to my house in Leafy Lane, where we sat on the back patio and toasted my success. Maybe I *would* get a telly, then I could put *The Generation Game* on for Mam while Dad and me stepped over to The Shepherds.

By the time I opened Bentinck's reception door, I was . . . whistling. Boston's 'More Than A Feeling', in time with the clack of my light-beige cowboy boots. I shook hands with the man and told him about rebuilding the Suzuki, but my mind was still on Leafy Lane, wondering whether I'd have remote-control gates. I heard him saying something about pit closures, the government, Margaret Thatcher and Arthur Scargill. And apprenticeships. That caught my attention.

What?

No more engineering apprenticeships that year. Margaret Thatcher, Arthur Scargill – he repeated the names – then told me about twenty-three uneconomical pits and the increased chances of strike action. If I fancied face work, no problem, but apprenticeships were on hold.

He sighed and shook his head. *'Tha can ay annuther goo next year.'*

And that was it.

I'd never actually had a job with the NCB and for a long time hadn't been sure I wanted one; but I'd been so sure of getting one that being told I wasn't getting one felt like . . . a swizz. Like I'd been sacked a second after signing my contract. What about my remote-control gates and the souped-up RS2000? What about King Coal? You summoned me. I came to Bentinck. I accepted a future that lined up with my dusty black destiny.

I rode home the long way. Up to the Countryman pub, turn left, along Pinxton Lane, right at the crossroads, up to the roundabout that joined up with the A38, down to the junction at Somercotes, up to the Pit Houses in Jubilee and back to Selston. Plenty of time to think about these people . . . Thatcher and Scargill. Two names that had suddenly burst through the doors of Bentinck and trampled all over my lifetime of unplanned planning.

Margaret Thatcher, elected as Britain's first female prime minister in 1979. Voted for by more than a few people in Selston . . . more than a few working-class people all over Britain. People who were fed up with the Labour Party, the stone-deaf union bigwigs and their socialist utopia.

And Arthur Scargill, 'left-wing firebrand', president of the Yorkshire Area of the National Union of Mineworkers and soon to be elected president of the National Union of Mineworkers. His 70 per cent share of the vote in that national election no doubt included a sizeable chunk of Nottinghamshire miners, but there was another sizeable chunk – in Selston and the surrounding villages – that had long regarded him as 'a bit of a cunt'. Even worse, he was seen as 'a communist cunt'.

Margaret Thatcher's success was built, in part, on a promise to tackle the union militancy that had plagued much of the '70s, and that ultimately led to confrontation with Arthur Scargill and the NUM. Which in turn led to me leaving Bentinck empty-handed. The battle between two political/ industrial heavyweights had ended here . . . with Selston's answer to Dennis Hopper.

The reaction at home was pretty much as I'd expected. My dad said nothing – he didn't even stop rolling his fag – and my mam set fire to the outside lavvy. Followed by, *'Y' mun gett thee sen a job on t' face then. Ah were in service at fourteen.'*

I explained that I wasn't afraid of the dirt or the danger of the coalface, but I wanted to make the best of my Clever Bugger credentials. I now had a plan. And I liked the idea of being an engineering apprentice.

'Oh, ah,' snorted my mam. *'An' 'ow yo goona gerra job as a 'prentice?'*

I mentioned the bloke who now worked at the main Ford garage in Derby. I listed the companies that offered appren- ticeships: British Telecom, British Gas, the Electricity Board, the new Kodak factory in Annesley, Ashfield Council, the

Army, the Royal Air Force. I threw in NASA and Suzuki, just for good measure.

Another snort. *'Yo . . . in t' army? They winna ha' yo. Tew much o' a know-it-all. That's yoar bluddy problem.'*

I tried explaining that the army – not to mention NASA and Suzuki – were keen on know-it-alls, but it was obvious that, like Mr Hall, I was wasting my time. The earth was flat, Michael Faraday worked on the meat counter at the Buxtons Hill Co-op and I was a fool.

I tried explaining that I could get the addresses for all these companies from the library and write letters to them, but it was obvious that I was wasting even more of my time. $F = mc^2$, Florence Nightingale was the new landlady at the Miners Arms on Inkerman Road and I . . . couldn't even be bothered.

Snort. *'Bluddy books an' wods. That's yoar answer t' everythin'.'*
Not this time, it wasn't.

My mam looked at me. And sometimes at the fridge. *'Tha's either gooin' bakk t' Bentinck and gerrin' a job or tha can bugger off an' live somewheer else.'*

I wanted to ask if she meant me or the fridge, but I didn't. I waited until my mam set fire to my dad and watched Dad's embers dancing their way up towards the darkest of dark skies; towards that hot, damp, dusty, pitch-black heaven. And then I left home.

As I still had a key to the back door, I wasn't technically homeless, more of an unwelcome guest. But I took her at her word and began a round of one- and two-night stands with various friends, as well as nights spent in a tent I kept hidden in the thickly forested area on the far side of the M1. I did try the House of Trees, but it was no longer there. Torn down, along with so many hedgerows and old barns.

Some of my 'homelessness' was all right. More than all right. One lad who had been in my class also happened to have a Suzuki AP50. We'd been out on a few rides together and when I explained what had happened, he asked his mam and dad if I could stay with them. I spent most Fridays,

Saturdays and Sundays at their house and, in return, I looked after his AP50 and helped his dad dig the garden. None of the family ever mentioned my mam, but I got the feeling they knew her and knew what she was like. I hoped they didn't feel sorry for me. Feeling sorry for me would have been a bit . . . much.

There was a kettle and sink at the garage and, as my evening shift meant I was the one to lock up, I spent several nights in the storeroom, sleeping on a layer of foam that I found down by Sperry Brook. One night was even spent at Wingfield Manor. The crypt underneath the main building was warm and had a vaulted ceiling. The farthest corner was dark on a summer's day; in the middle of the night, it was robbed of even the faintest flecks of light and colour. A strange realm, where darkness and gravity combined. A black hole, just a few miles west of Alfreton. I slept little and swear that I saw things. I was scared, but convinced myself that the overnight vigil was a continuance of my earlier hardcore physical and mental fitness routine. Not eating 99s, knife-throwing ventriloquism and facing ghosts.

I liked that. Liked the idea that I was facing down my own ghosts. The ghosts of my past, present and future. I wondered what they'd look like. The past? Great-Great-Great-Grandad Thomas. The present? A quivering ball of fire that floated randomly, hither and thither, loudly complaining about the price of fags.

As for the future? Maybe shave off the bumfluff beard? Wash my jeans? Write some letters? Make the best of my makeshift plan?

I'd read about posh students taking something called a 'gap year', interrailing around Europe or building water wells in Africa, and decided my predicament wasn't that much different. A gap couple-of-months-over-the-summer. With Kinshasa, Sydney, Paris and Madrid replaced by Codnor, Westwood, Somercotes and Ripley. And the library, where I spent whole days at the table by the records, replying to every promising job advert in every local paper. Rosemary let me use the phone to call the council and ask them about a grant that would allow me to enrol on a course at technical college and hopefully pick up a couple of HNC and BTEC qualifications.

Writing the job application letters was easy. I had a neat and steady hand, a thesaurus and a tasty set of exam results. Putting an address on those letters wasn't so easy. As romantic as 'The Crypt, Wingfield Manor, South Wingfield, Near Alfreton, Derbyshire, DE55 7NH' looked in the top corner of my letters, it didn't send out the right signals. There was no way they were going to employ the Earl of Shrewsbury or the chancellor of the Duchy of Lancaster.

So I asked the lad with the AP50 if his mam would mind me putting their address on the top right corner of my letters. She didn't mind at all and every couple of days I would knock on their door, have a cup of tea and read my letters. British Telecom sounded promising but the apprenticeship application process didn't start till later in the year. One letter was from a firm in Kirkby, asking me to come in for an interview the following week.

At the factory, a man who introduced himself as the technical director seemed fascinated by the story – as detailed in my letter – of me stripping and rebuilding my gleaming Suzuki motorbike. I showed him around the bike and he smiled as he asked me questions about gearboxes, hydraulics and electrical relays. Then another man took me to a small office and handed me several sheets of questions and mathematical problems – a sort of IQ / competence / thinking-out-of-the-box test. He asked me if I wanted a glass of water and said he'd be back in forty-five minutes.

In forty-five and a few minutes, I was told that I'd scored 93 per cent. The men seemed pleased. Laughing, the technical director said that my score had actually beaten the factory's chief engineer. They left me in the office for a few minutes more. I tried tugging on my beard, but I no longer had one. Eventually, the technical director walked in with another man, introduced as the managing director. We all sat down and the technical director told the managing director that I was a Clever Bugger. He actually used that phrase, which made me smile.

'This lad 'ere's a Clever Bugger.'

When I did finally head home at the end of those two months, it was to announce that I'd secured a place as an apprentice multi-disciplined engineer at a medium-sized injection moulding concern in Kirkby. Earning 35 quid a week. No more gravy train. My first year would be spent full-time at Chesterfield Technical College, learning electrical, mechanical and hydraulic engineering, welding, metalwork and maths. In one of the first maths exams, I scored 100 per cent. Not only that, but I'd arrived at one particular answer in a way that the lecturer had never come across. Steve Carlin, one of the other lads in the class, was sitting in the row in front of me. He was wearing a pair of jeans that I would buy off him a few weeks later and a chunky-knit turquoise V-neck jumper. When the test results were announced, he slowly turned his head to the left and looked at me.

'You wanker!'

During the college holidays, I went to work at the factory in Kirkby. I spent days and days rewiring this and that machine or sitting through this and that practical assessment. I sailed through most of them, enjoying the patience and precision required to build power supplies and repair things.

One of the electrical lecturers at Chesterfield was forever reminding us that, where engineering is concerned, 'Neatness is fifty per cent of the job.' I agreed wholeheartedly. That was why my once ramshackle AP50 looked better than most of the others in Selston.

The technical director took a shine to me. He invited me to join him on a visit to a big factory in Sheffield where we listened to a talk about the latest injection moulding machines . . . bigger than the ones we had and digitally controlled. On the way back, sitting in his dark blue two-litre Mk4 Ford Cortina, we chatted about how, even though these new machines were far more complex than anything we'd seen before, they were much easier to maintain. Instead of having to work out which resistor or relay had blown on which circuit board, they came with a series of error codes that told you where the fault was. All you had to do was unplug the offending circuit board and slot in a new one.

On the M1, heading towards Junction 27, he started swearing and cracking jokes. He talked about when he had a BSA Gold Star in the 1960s. I immediately asked him if it was the 350cc or 500cc model. I wanted him to know that I knew what he was talking about. We talked about oil leaks and he decided to see how fast his two-litre Cortina would go. At 115 mph, we agreed that we'd gone fast enough.

When we stopped to fill up for petrol, he bought me a cup of coffee. My first cup of coffee. At the factory, he drank a lot of coffee and I decided that I would drink coffee, too. I bought a jar of Nescafé. The good stuff, as advertised by Gareth Hunt, who used to be in *The New Avengers* with Joanna Lumley.

They use three types of coffee bean, you know.

Smack Bang in the Middle of the Largest Industrial Dispute in British Post-War History

Despite the invite, I never went back to see the man at Bentinck. Even if I hadn't got the job in Kirkby, there wasn't much point. During the following couple of years, 1982 and 1983, the government continued to push for pit closures. Understandably, the NUM and Arthur 'left-wing firebrand' Scargill continued to push for zero pit closures. Thousands of real jobs were at stake and few pits were interested in taking on any wet-behind-the-ears apprentices.

Talk at the time was centred on the vile Tory government of Margaret Thatcher and how she was hell-bent on murdering Britain's mining industry. Even I knew that the situation couldn't be explained in such simple terms. On telly, Arthur Scargill was complaining about Margaret Thatcher closing all the pits, but he never mentioned Harold Wilson and the 253 pits that had been closed during his two terms in office. He never brought up the 1970s speeches by Tony Benn – the political left's favourite MP – that talked about closing uneconomical pits.

The truth was – as perhaps Dad and me had realised during the turbulent elections of 1974 – that mining had been in trouble for years. From a fairly solid figure of around 750,000 in the late '50s, the number of men employed by British mines had fallen to around 250,000 in the early '70s. There were other factors, too, like the dwindling post-war need for coal and the introduction of mechanisation – just over 5 per cent of coal output was mechanised in the late '50s, rising to over 90 per cent in the early '70s. And the arrival of cheap coal from Eastern Europe, courtesy of the new globalised marketplace.

Something had to give.

The ensuing miners' strike in 1984, while not a complete surprise, still had the power to shock. In mining communities, strikes tend to have a build-up. Like thunderstorms, we know when they're coming. Wet fingers were poked skywards as we felt that early spring air – a fall in barometric pressure, an exponential increase in brooding anticipation and a grim, leaden sky. The colour of hateful mistrust from both sides.

When the strike eventually started on 6 March, all the action took place at Cortonwood, a South Yorkshire mine that was due for closure. Selston carried on as normal. Normal-ish. Yes, people argued in pubs about unions and collective action; people expected upheaval. But those same people still believed that mining was Britain's most enduring industry. Especially in the East Midlands, which was home to some of the most profitable pits in the country. Even during the strike, the idea of a Selston without mining was too outlandish to contemplate. It would never happen.

When Arthur Scargill attempted to turn the Cortonwood dispute into a national issue and called for an all-out strike, most of the Nottinghamshire miners – and they weren't alone – refused to down tools. They pointed out that the NUM hadn't called a national ballot over strike action. In other words, the union leaders hadn't bothered to ask the miners if they wanted to go on strike; they simply ordered them to go on strike.

Thankfully, Dad had taken early retirement in 1983, the year before it all started. He never talked about the strike, but I assumed that he'd known what was coming and thought his time would be better spent planting vegetables or collecting dandelion leaves down the Dumbles. To be honest, I couldn't imagine him or many of his mates joining the picket lines, throwing bricks at bus windows or jabbing police horses with broom handles.

Even as the strike gathered momentum, there were still about 30,000 miners who continued to clock on for their shifts, and the majority of them were in Nottinghamshire, Derbyshire and Leicestershire. At the Co-op, I saw a former classmate who was now working on the face at Bentinck. He told me he was earning almost as much money as his dad, had bought his first car and had no intention of going on strike. His parting words were reserved for Arthur Scargill.

'*That cunt shudd 'a 'ad a ballot. Ay shit it 'cos ay noo ay'd lose.*'

As well as sitting smack bang on the boundary between Nottinghamshire and Derbyshire, smack bang on the edge of the Erewash Valley and smack bang on the border of Nottinghamshire's exposed coalfield (close to the surface) and its older, concealed coalfield (much deeper), Selston was now sitting smack bang in the middle of the largest industrial dispute in British post-war history. A lot of Selston's, Underwood's, Brinsley's, Jacksdale's, Kirkby's and Annesley's miners wanted to work. The ones that had decided to strike were joined by other miners, arriving from places like Yorkshire, Durham, Scotland, Kent and Wales. Together, they were intent on stopping any local miners from getting to work. Hundreds of police were also arriving from places like Yorkshire, Durham, Scotland, Kent and Wales, intent on making sure those local miners managed to get to work.

And that's when things started to turn nasty.

I was riding to work one morning on my gleaming Suzuki motorbike when I was stopped by half a dozen men who were standing in the middle of the road at the start of Bentinck Lane. As I pulled up, one of them reached for my keys, turned off the bike and chucked the keys into the grass at the side of the road.

'*Off t' wokk then?*' another one asked in an accent that was much more northern than I was used to.

'Ah.'

'A' tha at Bentinck?'

'Ah'm norr a miner.'

'Oh, aah. Wot's them then?' He pointed at my gloves.

Every now and then, Dad used to bring stuff back from the pit. Soap, rolls of cable and wire, boots, donkey jackets, overalls and gloves – a particular type of thick yellow glove that was instantly recognisable as pit property. I sometimes used a pair as my summer motorbike gloves.

I tried to explain that my dad had nicked the gloves, but that made things worse. So my dad was also a Nottinghamshire miner. One of the men pushed me off my bike. I fell, smacking my helmet on the road and trapping my leg under the hot exhaust. By the time I limped my gleaming Suzuki motorbike back home, it wasn't quite as gleaming. The petrol tank had a rock-sized dent, the back tyre had a Stanley-knife-sized hole and I still had no keys.

The following morning, I borrowed my mate's pushbike. I approached Bentinck Lane slowly, making sure the coast was clear. Under normal circumstances, it was the most exquisite section of my journey to work. After passing over the M1 between Junctions 27 and 28, the road dips steeply down, winding between fields and tall hedges. Kirkby is visible in the distance, but the only building that ever caught my eye was the roof and chimneys of Bentinck Colliery: another shining example of coal and nature working together. The sweep of the road, the line of the horizon, the various angles of the hills and fields all drew your eye to the colliery. From the top of the lane, the distance and height difference gave it the appearance of a children's toy. A farm or factory that a five-year-old would festoon with plastic cows and Matchbox lorries.

One previous morning, just as the sun was coming up, I took a wonderful picture of Bentinck Colliery. It was a summer Saturday and I set off for work at about five thirty. From our house, the eastern horizon was blocked by hills, but there were certain sunrises that managed to shine up and over those hills, sprinkling their polychromatic fairy dust all over the estate. As soon as I looked out of my bedroom window, I knew this was one such sunrise and shoved my cranky Zenit EM camera into my rucksack.

Somewhere close to where those blokes removed the gleam from my gleaming Suzuki motorbike, I had stopped and taken one picture of the grey colliery framed by a purple sunrise sky. Mr Finch, the local chemist, developed the film and when I went to pick it up, he mentioned a purple picture that he liked. I flicked through the twenty-four images and pulled out the purple one.

Although me and Mr Finch the chemist both lived in Selston, we were on different sides of the Great Divide. In all my years as a customer, I don't think I'd said much more to him than, *'Large crepe bandage, please.'* But there we were, leaning over his neat and tidy counter, nodding our heads at a photograph I had taken.

'What camera have you got?' asked Mr Finch the chemist.

'Zenit EM. Ah gorrit frumm that second-'and shop in Af'ton. It were ownee ten quid. Shutter were stuck oppen burr ah fixed it.'

Mr Finch the chemist said it wasn't easy to make a pit look beautiful. Although I disagreed with him, I kept my disagreement to myself. I didn't want to ruin the moment.

A simple photo had narrowed the Great Divide. From then on, every time I went into his shop, Mr Finch the chemist would say hello and ask me how I was.

'Hello, Benny. How are you keeping?'

I didn't want to ruin those moments either. And 'Benny' wasn't so bad.

On the morning with the pushbike, the sky was grey, not purple. Bentinck's chimneys were barely visible through the mist and I could hear faint shouting from somewhere across the fields. As I rode over the railway tracks at the bottom of the lane and started pedalling towards the pit entrance, I saw what people were shouting about.

On the left-hand side of the road was a large group of striking miners and flying pickets. They were waving fists and two-fingered salutes. And shouting. Throwing bricks and bottles across the road. And shouting. On the right-hand side of the road, a long line of policemen was guarding the entrance to the pit, occasionally making way for a car or minibus taking non-striking miners to work. There were a few policemen on horses, too.

Sometimes, a small group of striking miners would tear across the road, squaring up to the policemen, gesturing at them like kids in the playground.

'C'mon then! Ah'll smash y' face in! Gi' us y' dinner money!'

I climbed off my bike and watched, wishing I'd brought my Zenit EM. A coach came down the road from Kirkby and this seemed to rile the picketing miners. The noise rose accordingly. One of the men rushed into the road, standing menacingly in front of the coach. Several more followed him. Then all the men spilled into the road. And all the policemen spilled into the road.

I half ran/half walked, pushing the bike past the melee and up towards the Countryman pub. There were a handful of TV cameras dotted about, eager to capture any bloodshed, and I've often wondered if one of them also captured me and my mate's BSA Javelin sneaking past.

Police patrols in Selston were a regular sight over that summer. I was stopped and manhandled multiple times, but as soon as they saw the address on my driving licence and worked out I wasn't a miner, they calmed down.

'There's been some trouble,' they would tell me by way of an explanation. And they weren't just talking about the scuffles at Bentinck. Groups of men that weren't from Selston had been seen walking and driving around the village, waiting outside the houses of working miners, damaging cars and spray-painting walls with 'Scab Bastard'. Men we'd never seen before. Late at night. Watching. A brick through a window here; some dog shit shoved through a letter box there.

I saw some of these men walk past our front gate one evening. I was in my bedroom and my dad was digging the front garden. One of the men opened the gate and stood on the steps, looking straight at my dad. Dad stood up and stabbed the spade down hard into the soil between them. By the time I got downstairs and out of the front door, the man was closing the gate behind him and the three of them headed down the street. Dad carried on digging. But faster. I'd never seen anyone dig with such distaste and anger before. The whole border at the top of the lawn, finished in a matter of minutes.

The strike lasted 362 days, three days short of a year. And it was an odd 362 days. At times, so charged with bleak, poisonous emotion that it felt like we were living in one of those dystopian, near-future dramas on the BBC. Of course, we saw the news stories on telly about families in Nottinghamshire and other counties relying on food banks and the kindness of strangers, and, yes, we felt sorry for them. But our sympathy was tempered by those bricks that had been thrown through our windows and the dog shit that had been shoved through our letter boxes. Tempered, too, by the look on my dad's face when that man opened our front gate. Tempered by the bloodstain on the pavement where that Nottinghamshire miner's wife was attacked as she walked her kids to school.

Tempered by the death of David Wilkie, a taxi driver from Mid Glamorgan who was taking a non-striking miner to work. Wilkie was a thirty-five-year-old father of three and his fiancée gave birth to a fourth child just six weeks later. A man simply doing his job, but two striking miners decided to drop a concrete block from a footbridge onto his car. I wondered if those two miners were thinking about the NUM motto when they let go of that 50-pound block of concrete: 'The past we inherit, the future we build.'

And after those 362 days? What kind of future was built? Did the miners' strike affect Selston? Did it affect me, my mam and dad? Despite all the violence and the rhetoric, the end of the strike just fizzled out with a disappointing phutt. The long-term repercussions, though, were neither fizzled nor phutt-ish. Physically, there was little difference, apart from

Dad's front garden looking particularly smart. It wasn't as if all the miners and all the pits suddenly disappeared. But, over time, the changes started happening and they kept on happening. Jobs were lost, people moved away, shops closed, pubs closed and, yes, even pits closed.

Work at Pye Hill Pit No. 1 and 2 in Underwood and Jacksdale ended in 1985. At Moorgreen in 1985. At Hucknall No. 2 in 1986. At Newstead in 1987. At Mansfield in 1988. And so on.

The unthinkable had been thunk.

In the aftermath of the strike, Yorkshire and Durham miners complained about the way they'd been treated by the police. They complained, too, about the Nottinghamshire miners, Margaret Thatcher and capitalism. But I never heard one of them explain why they had begun attacking Nottinghamshire miners' wives while they were walking their kids to school. They didn't even seem that keen on saying sorry for attacking those miners' wives while they were walking their kids to school. Even Margaret Thatcher − the archvillainess of the whole piece − had entered into 'cordial' correspondence with wives of working miners. The NUM and the striking mining families − who in some cases lived in the same village, shared the same school and were even related to the women who were being attacked − simply continued to attack them.

For their part, the Nottinghamshire miners − and I found it hard not to agree with them − insisted that they shouldn't be held responsible for the collapse of British mining. Bringing up the strike in conversation could be dangerous. But if it was mentioned, most Nottinghamshire miners − and again, I found it hard not to agree − reminded everyone of all the pits that had already closed in the '50s and '60s, many under Labour governments that had no connection with the strike or Margaret Thatcher. They reminded everyone of the increase in mechanised mining. They reminded everyone that, because no ballot had been held, the strike was undemocratic. They drew attention to recent, previous national ballots that had resulted in Britain's miners voting against industrial action. Irrespective of what Arthur Scargill thought and felt, said the Nottinghamshire miners, the majority of men he was supposed to represent did not want to go on strike.

Did those Yorkshire and Durham miners honestly believe that their Nottinghamshire colleagues refused to go on strike because they wanted to annihilate the mining industry? Because they wanted to destroy communities and break up families? Because they wanted wives to stop speaking to husbands? Fathers to stop speaking to sons? Friends and brothers to stop speaking to friends and brothers? One lad I knew stopped speaking to his dad because his dad wouldn't come out on strike. In the end, they both lost their jobs, and the son moved his family up north somewhere. Even after his dad died, the lad stayed up north somewhere.

I wondered what Great-Great-Great-Grandad Thomas would have made of it all. I wondered what would have happened if, as that corf was hurtling towards his head, the almighty gods of mining had given him a quick glimpse of 1984 and 1985. It wasn't just his own life flashing before his eyes . . . it was the lives of every miner from every age.

Did he have to die for this?

Although it was the strike and the uncertainty of the previous three or four years that prevented me going down a mine, I couldn't work out if I was supposed to be bitter or thankful. Bitter at mining's wretched demise or thankful that I never had to go through what my dad went through? If bitterness took the spoils, whom should I blame? Margaret Thatcher? Arthur Scargill? Globalisation and the arrival of cheap coal from Poland? And if thankfulness took the spoils, who should I thank? Thatcher and Scargill? Dad and the Nottinghamshire miners? Harold Wilson and those 253 closed pits?

Two questions . . . with (effectively) the same answer.

There was a part of me that resented the strike. Resented it for taking away my birthright. My chance to do what was expected of me. To not let my dad down. Just my dad . . . I wasn't worried about my mam. As things turned out, it's unlikely Dad would have cared either way. By the time the strike ended, he was too busy trying to catch his breath as he walked up the hill to the post office. Too busy pretending he wasn't going to die.

In the end, it became just another of those questions I'd never asked him. Dad, did you really want me to get a job at the pit? Did you really hope and expect that I would follow you into the Mouth of Hell?

And what if I had ever asked him that question? What answer would I have got? He would have continued reading his newspaper, sanding down the nicotine on his fingers with a matchbox and chomping on a single jelly bean left over from Christmas.

But he would have said . . . nothing.

And We Laughed . . . Like Seagulls

Never one to make a fuss about anything, Dad treated retirement like any other normal day. He did have one complaint, though. And while it must have hurt him deeply, he mentioned it only once . . . on his first morning of freedom. Mentioned it only in passing when I asked if he'd been presented with the ceremonial 'Gold Watch'.

Yes, Dad told me, his manager had presented him with a Gold Watch. And 500 quid. His reward for forty years down the pit was a Gold Watch and 500 quid. For a life's work. For risking that life. Over those forty years, 500 quid works out at about 5 pence a day. Dad and thousands upon thousands of other men just like him were paid 5 pence a day to die.

Dad kept the 500 quid, but threw the Gold Watch back in his manager's face.

Knowing that helped me understand why Dad was against Mam's idea for a retirement party. But it wasn't just the Gold Watch that had soured the occasion. Dad didn't really like parties. Although he enjoyed chatting, drinking and selling bingo tickets at the Tin Hat, he saw no reason to enjoy himself at home. Not in that way. Crisps in a bowl; tinned salmon on a cob. All those people standing around, trying to think of things to say.

That never happened at the Tin Hat. There was no room for uncomfortable silences when you were selling bingo tickets. *Just a cupple, Norm.'*

"Ow do, Harry?'
'Middlin'.'
'Sih thee.'
'Aah . . . sih thee.'

In a sense, though, my mam was right. This was a cause for some sort of celebration. After more than forty years underground, retirement was a chance to do all those things that Dad had wanted to do and see all those places he'd wanted to see. Unfortunately, much of Mam's enthusiasm came from the fact that retirement meant Dad would get up and go to bed at the same time as everyone else. For her, this was the equivalent of getting a live-in butler who could also supply regular updates on the weather and the arrival of new bus stops.

If there were any things he wanted to do or places he wanted to see, they were quickly forgotten. His days were now filled with a long list of jobs that Mam had lined up for him. Dad was no slouch on the coalface, but without the impetus provided by pit banter and the constant fear of death, his work rate began to tail off. Although he did manage to sort out the clothes line and decorate the front room, each job took days, weeks or even months to complete. It wasn't unusual to walk into the front room and find him slumped on the sofa, roll of wallpaper in one hand, scissors in the other, snoring happily.

If he was my gleaming Suzuki motorbike, I would have given Dad a bit of a freshen up. Changed the oil, air filter, spark plug and so on. All the stuff that gives tired old machines a new lease of life. As it was, I was forced to watch this tired old machine struggle to get started in the morning. I watched it change shape. The lean, muscular torso that had once paraded down Skegness seafront turned soft and flabby. The lean, muscular arms and legs that had raced across the Rezzer grew thin and withered. Pipe-cleaner arms and legs on a roly-poly Plasticine body.

Instead of guiding Dad towards roads never travelled and adventures unimagined, retirement attacked his body like a slow-acting poison. By the autumn of 1987, just four years after his 60th birthday retirement, he'd stopped shaving and bathing. His hair was no longer combed and styled. He wore the same clothes week in, week out. He spent more of the day and night in bed. His hands refused to grip. The change in body shape was caused by water retention and the water retention was caused by his lungs, which no longer worked properly.

At first, he fought hard against his lungs, but they were just too strong . . . or weak, depending on which way you looked at it.

'Ah'm jiggered,' my dad would tell anyone and no one. 'Raight jiggered.'

Dad wasn't the only one who was jiggered. Many of his Tin Hat friends and mining colleagues lived on the estate and most of them ended up like Dad. Every now and then, we'd hear about another one of them going into hospital or worse. That must have affected Dad. He got slower and sleepier. I wanted to move his bed downstairs, but he said that he'd rather sleep on the floor in his bedroom. I'm not sure how, but he managed to get himself back up those thirteen stairs every night: one arm, rest; one knee, rest; other arm, rest; other knee, much longer rest.

And start again.

No breaking wind, this time.

Like our old evenings in the Pit Houses, evenings on the estate tended to be conducted in the dark. Mam was still blind, so she didn't need the lights on. Dad would sit in the back room, listening to the radio and smoking the occasional fag, so he didn't need the lights on. I had a small lamp in my bedroom, but hardly ever turned on a main light because it made the world seem so painfully stark. Like suddenly training a spotlight on an escaping prisoner, blinking and suspended halfway up a 20-foot fence.

Some nights, I would sit with Dad in the dark, in the back room, listening to the radio or playing records on the now well-worn Fidelity record player. Apart from the music of

Hank Williams or Tex Ritter, we and the back room were silent. And when the music stopped, all we could hear was the overflow from Mam and the telly in the front room. The theme tune to *Coronation Street* or *Bergerac*. Adverts for British Telecom or Hamlet cigars.

The back room faced almost south and west, which meant that Dad and me got the best of evening's colours. As I sat on my chair in the opposite corner, I could see his back, shoulders and head silhouetted against the window, against the sky. At first, I was able to pick out little details like the shape of his jowly face or the number of shirt buttons that were undone, but as the clock ticked and the colours fled, Dad would gently disintegrate.

At times, he forgot I was there and would start prodding and poking at his chest. Or softly pressing his stomach with both hands. Or quietly coughing. If I spoke or moved, he would suddenly sit up straight and take a swig of tea.

'*A*' tha raight?' I'd ask.

'*Champion*,' said Dad.

Eventually, we'd hear Mam switch off the telly and start getting ready for bed. Dad and me would let whatever song was playing get to the end, then he'd turn off the radio or I'd turn off the record player and . . . nothing. I'd sit back down on my chair. Dad would sit on his chair. Some evenings, he obviously felt the cold and pulled on a cardigan and a pair of grey gloves. The gloves came with their own sound: the deep *doi-nnng* as Mam walked into the kitchen door, on her way to make a cup of tea to take upstairs. They came with their own smell: fags and Dad's body odour. Their own mirror image: moonlight glinting on the glass doors of the wall unit, bright enough to show the reflection of Dad's movements and several shelves laden with nobby ornaments and knick-knacks.

The gloves seemed to increase the pressure. The overwhelming pressure of all that air in the back room. So much air and yet so little of it was of any use to Dad.

There was a brief period where I brought down my

bedroom lamp and read stuff to Dad. Stories in the paper, even the odd book. Dad had half a dozen paperbacks, propped up in a cupboard, in the kitchen, next to tins and tubs of screws and nails. There was the Dennis Wheatley he used to read in the Pit House and a couple of similarly black-magic-themed tales. Not quite what I was after.

What about Zane Grey? *Riders of the Purple Sage.* The cover showed a rough picture of a weary cowboy and his lady friend . . . in front of a sad, beautiful sky. And the words: 'One of the most famous of all Western novels!' That sounded more like it.

A sharp clip-clop of iron-shod hoofs deadened and died away, and clouds of yellow dust drifted from under the cottonwoods out over the sage.

I did my best to read well; steady and clear. I wanted Dad to get the full effect of the hooves and the yellow dust. Sometimes, he would turn to the window and stare. I hoped he was staring out on to something other than Mr Winton's pigeon coop. A wide-open Dakotan prairie, perhaps. The Blue Ridge Mountains of Virginia. Or the riders and their purple sage.

And that gave me an idea.

I borrowed the wheelchair off a bloke at work – he'd been given one by King's Mill Hospital in Mansfield when he broke his leg, but had never taken it back. It was large and sturdy; not quite an off-roader, but more than capable of coping with a few bumps and bangs. When I told Dad about my idea, he screwed up his face and shook his head.

'Ah'm non bothered.'

I played it down. It was nothing special. If the weather wasn't too bad, I was going down the Dumbles at the weekend. Did he want to come with me?

'Ah'm non bothered.'

It'd be nice to have a drink at The Shepherds.

'*Ah'm non bothered.*'

Or the Dixies.

'*Ah'm non bothered.*'

So I left it at that.

But on Saturday morning, Dad was up earlier than usual. I heard him coughing and moving things in his bedroom, then he walked across the landing to my bedroom door. His hair was combed and styled, and the grey bristles had disappeared from his chin.

'*Ah dunna want ennyboddy knowin'.*' He cocked his thumb towards Mam's bedroom.

I pointed out that he couldn't just disappear for half the day. What would we tell her?

'*Tell 'er nowt.*'

That's exactly how it happened. I fetched the wheelchair out of the coal shed, stood it by the back door and waited for Dad. The weather was warm, as I'd hoped, but Dad still brought a jacket. We left the back door open, neither of us saying a word to my mam as I pushed Dad round to the front of the house and up the steps. Even with the ballooning torso, his weight had dropped dramatically and getting him onto the road was much easier than I'd expected.

We both knew that news travelled fast on the estate and it wouldn't take long for someone to tell Mam what they'd seen. Several pairs of eyes glowered out of net curtains as we turned the corner at the bottom of our road. One of the owners of the inquisitive eyes would no doubt enjoy popping over to see my mam.

'*Ay-upp, Hilda. Ah've just seen Norm in a wheelchair. Is ay in a wheelchair then? Is ay disabled?*'

I would have loved to hear what she said to the kindly messenger.

'*Disabled? Ay's non disabled. Winna gerrupp in a mornin', that's wot's wrong wi' 'im. Ay's upp theer, nah. Fast-on.*'

The wheelchair came with brakes – like you get on a bike – and footrests at the back that helped tilt the chair when you were mounting pavements. If I stood on the footrests and kept my weight forward, the chair carried us both. Controlling it wasn't easy, but with a separate brake on each wheel, I could just about guide it into corners.

By the time we reached the jitty, halfway down the estate, we had picked up some speed.

'*A' tha raight?*' I asked.

'*Champion,*' said Dad, the breeze juggling his rejuvenated grey hair. Then I heard him shout. '*A' tha raight, Billy?*'

Billy Gee was in his garden and the look on his face told me that he wasn't quite sure who – or, indeed, what – it was that he'd seen zooming down the pavement. Then he twigged.

'*A' tha raight, Norm?*' he shouted back. There was no hint of surprise or concern in Billy's greeting. It was as tonally characterless as if he and Dad had just passed each other in the lavvy at the Tin Hat.

The road levelled out after the jitty, but momentum carried us all the way to the bottom of the estate. The stretch of wasteland that had provided camouflage for so many games of army (groups of lads making machine-gun noises at each other) had been replaced by a Kwik Save supermarket and car park.

'*When they oppened it, may an' y' mam were t' fost tew in theer,*' he said waving at the Kwik Save. '*They sedd they'd gi' us a chicken. They nivver gen it us. An' it were rainin'.*'

'*Wot d' yuh mean? Gi' yuh a chicken?*' I wondered.

'*They 'ad a sign upp. Fost tew in theer gorra free chicken. They nivver gen it us.*'

As a metaphor for Dad's life, it was clumsy, but it did the job. He'd done everything that had been asked of him. Worked hard, paid his way and pulled his weight. He'd even queued outside Kwik Save in the rain . . . but there was no free chicken.

When he could still go shopping, Dad went to Kwik Save, but he didn't enjoy it. He only went because my mam liked going there. My mam liked going there because she could hold on to the handle of the shopping trolley while Dad guided it up and down the aisles. Mam was convinced that anyone looking at her would assume she was the one who was pushing it. They would assume she could see where she was pushing it. With her 20/20 vision.

Dad had always preferred shopping at one of the four Co-ops in Selston, but there was no longer any need for four Co-ops now that we had a Kwik Save. More people were also shopping further afield. Families would jump in their cars and drive to even bigger supermarkets in Alfreton and Sutton every Saturday. One by one, the four Co-ops closed. Mr Finch the chemist closed his chemist shop. Most of the old corner shops – the ones that had sold everything from Trebor mints to caustic soda – became houses. Butchers and barbers and grocers and betting shops went the same way. Even the cobbler, a marvellous and peculiar old man in a marvellous and peculiar brown apron, decorated with several decades' worth of dye and glue that gave it the appearance of an exotic animal skin.

All of that and more. Lost for ever.

Me, Dad and the wheelchair took the usual route to the Dumbles. The way we used to take when Dad let me carry his air rifle. After crossing the main road by the Bottom Rec, we rolled towards the cinder track that marked the entrance. From somewhere in an almost forgotten past, Dad suddenly produced a story.

'*Can yuh remember when way tukk yuh t' t' doctors? Y' mam thought there wa' summat wrong wi' yuh. Sedd shay cun't understand wot yuh were sayin' half o' t' time. Rekkoned yuh were talkin' foreign. Shay thought it were 'cos y' dad were foreign.*'

'*Wot did t' doctor say?*' I wanted to know.

Dad shook his head. '*Nowt.*'

'*Weer wa' me dad frumm?*' I wanted to know that as well.

Dad shook his head. '*Dunna know. It were Russia . . . o' somewhere.*'

It wasn't exactly a three-volume biography, but it was a start. Was it the start of his story? My story? Our story? I tried moving things along by asking him various questions about my life and his life, but there was obviously a lot he didn't want to talk about. Or simply couldn't talk about for one reason or another. The more I pushed, the less he said.

Nettles were OK, though. So we discussed nettles for a while. The lemony-like scent of the flowers, their excellent composting qualities and using the sting to treat arthritis. He wondered if there was still a big patch of nettles surrounding the trees by the bridge where we used to practise with his air rifle. There wasn't. In fact, most of the Dumbles' once wild hedgerows and unkempt edges of fields were much less wild and unkempt. Any messiness was of a modern kind: fag packets, johnnies, plastic bags and drinks cans.

There was still plenty of goosegrass and dandelion leaves to be gathered, but the thought of biting down on a bit of pink rubber or a fag end didn't do much for my appetite. As we paused by the bridge and its lack of nettles, Dad took the opportunity to contentedly roll a quick fag and smoked it . . . contentedly. Well, the bits in between the coughing seemed quite content.

Was he thinking about the time I took the air rifle all the way over to the fence and announced that I was going to land a bullseye? How long ago was that? Ten years ago? More? I was thinking about it. I could clearly see Dad, half laughing and half shouting: *'Tha'll nivver gerrit.'*

I did get it. I gorrit.

'Well, bugger may,' he said, tearing the target from the tree trunk. *'Well dunn . . . son!'*

That phrase was something he used when he was surprised and pleased. He'd say 'Well done', then leave a long pause before adding 'son'. He said it the time when he'd forgotten the bus fare to Nottingham. We were standing at the bus stop on Portland Road. The bus was due in a few minutes, so he reached into his pocket for the two pound notes that were unfortunately still tucked under the fruit bowl that sat on the dining table in the back room – the one with four chairs. We could have just waited another hour for the next bus, but I was excited and wanted to go to Nottingham. So I ran home to get the two pound notes.

Going back to the house was easy – no more than 300 yards, all downhill – but getting back to the bus stop took everything that my twelve-year-old body had to offer. Stomp-stomp-stomp up the hill, legs burning, flares flapping. I tried to think of a song that would match the rhythm of my running. 'Waterloo' by ABBA did the trick. As I sang along in my head, I imagined myself on the dance floor at the Tin Hat and knew that if I could carry on running to the end of the song, I would be at the bus stop.

Dad watched my head rise above the crest of the hill just as the bus drew up. I could see him speaking to the driver and pointing at me.

'*My, my,*' he explained to the bus driver, '*I tried to hold him back, but he was stronger.*'

'Oh, yeah,' said the bus driver with a smile, '*and now it seems your only chance is giving up the fight.*'

'*And how could I ever refuse?*' my dad wondered.

The bus driver was adamant. '*You feel like you win when you looooooossse!*'

Did he remember how fast I had run? How happy I had been? '*Dad, can yuh remember when yuh forgot t' bus fare t' Nottin'ham?*'

'*Aah.*' He nodded and just about managed to smile. '*Well dunn . . . son!*'

The Dumbles had once been Dad's second home and he'd walked its paths as if they were his own. He belonged there as much as the crows that scavenged in the fields and the kingfishers that flitted across the brook. As much as the midges and dragonflies. As much as the elderflower trees that had muscled into every hedgerow and the rosebay willowherb that gathered by the roadside fences.

In my early teens, I'd read Kathleen Basford's book, *The Green Man.* Basford explored a link between the iconic image of the Green Man – ancient, bearded face, garlanded by leaves and twigs – and the medieval understanding of nature. Lots of people in the papers and on the telly were talking about nature. Humanity's relationship with and understanding of nature's power was a constant reference point for books in the bestseller lists (*Jonathan Livingston Seagull*, *Watership Down*, the work of James Herriot, everybody was reading Tolkien), politics (the recent arrival of the Green Party), films (*The Blood on Satan's Claw* and *The Wicker Man*) and even children's telly dramas (the eerie folk-horror themes of *Children of the Stones*, in which the energy of a megalithic stone circle turns the local villagers into those very same megalithic stones).

Some of the pictures in Basford's book reminded me of my dad. Admittedly, Dad didn't have leaves and twigs growing out of his head, but if he'd ever woken up with leaves and twigs growing out of his head, I wouldn't have been surprised. Like Dad, the Green Man had a big nose and sad eyes. A face that had seen so much, but a face that would tell you nothing, no matter how many questions you asked.

Dad would have made a first-rate leader of this Johnny-come-lately green/ecological movement. Even without all the leaves and twigs growing out of his head, he was well qualified for the job. Never owned a car; walked or took the bus everywhere; foraged and recycled before most people knew the meaning of either word; made do and mended (some of his socks had lasted him fifteen or twenty years). Obviously, they would never make him leader because he'd worked at the pit.

He could have been a character in one of Tolkien's books, though. Here, surrounded by the rugged greenery and softly rising hills of the Dumbles, I could easily imagine Dad striding out o'er the Shire towards Bagthorpe and The Shepherds Rest – suitably Tolkienesque names.

He could be . . . Norman Coalhewer. Spending his days exchanging merry tales with the Elves of Rivendell.

'*A' tha raight, Elrond?*'

'*Non bad, Norm.*'

The Dumbles were hard-going with the wheelchair. After the little bridge without nettles, we started following

Bagthorpe Brook towards the road, but we ended up being forced further south, coming out almost opposite the medieval Wansley Hall.

'*Me dad wokked theer.*' Dad pointed to the remains of the hall.

'*Doin' wot?*'

'*Wokkin'. On t' farm.*' A minute or two later, he added: '*Summ days, ay were wokkin' wi' Wotsisname Lawrence. That book fella.*'

I hadn't expected that one. '*D. H. Lawrence? Wot worr 'e like?*'

'*Ah dunna know.*'

'*Din't yuh ask y' dad?*'

He shook his head. '*Wot for?*'

Was I at last beginning to understand why Dad didn't tell me much about his life? I tried to imagine the very moment my grandad told Dad about working at Wansley with D. H. Lawrence. Did Dad ask any questions about Wansley or Lawrence? Why would he? Although I might be desperate for stories about Grandad Jack sitting atop a giant haystack, gnawing on a chunk of home-made bread with one of England's greatest writers, my dad had all the information he needed. His dad used to work on a farm. With some Clever Bugger who had a beard. What was the Clever Bugger like? Don't know. Do you think he based one of his characters on Grandad Jack? Haven't got a clue.

Maybe Dad had the right idea when it came to questions and details like that. They make no difference. It is what it is, and we are where we are. Move along, now. Nothing to see here.

The Shepherds Rest was open but crowded. We tried the Dixies. Same story. Instead of heading up and across the fields as we would have done in the past, I decided to stay on the road. The wheelchair was beginning to sound squeaky, and I wasn't sure how many more humps and hollows it would manage. The road took us past my old school, Bagthorpe. Did Dad remember the eleven-plus? Was there any point in mentioning it? Would it have made any difference?

It is what it is, and we are where we are. Move along, now. Nothing to see here.

The home stretch took us up the long, steep and relentless Middlebrook Hill. I had to stop several times and stopped again when we finally got to the Top Rec. I parked the wheelchair and lay down on the grass beside it. Dad had set his jaw towards our estate but seemed to be scanning every inch of Selston. The new bits that had extended the village: groups of fancy houses where pubs used to be. The old bits: the chippy and the church.

He made a dismissive sound. *'There's more on it, burrit feels . . . less.'*

I asked him if he meant what I thought he meant. *'It's bigger, burrit feels less like Selston?'*

'Dunna know worramm lukkin' at.'

Pictures and maps from my dad's younger years showed lots of pit workings and lots of chimneys in Selston. Brick houses painted black by smog and soot. Although there were areas of farmland as you looked towards Jacksdale and dense woods over by Felley, much of Selston had been dug up or dumped on. A giant pit with a few houses dotted in between the various mineshafts and workshops.

Most people looking at those pictures would call the landscape ugly. I disagreed. Harsh, yes, but not ugly. I'd seen the beauty of it as a child – trailing in my dad's footsteps, walking past lapwings and gas bottles, swimming in the Rezzer – and, as I stood there with my dad, I could still see the beauty. Dad saw it, too. Just as he'd seen it as a boy; seen it as a young man. But the beauty was fading and Dad resented that. He missed what his home used to be.

Yes, the new Selston was cleaner and you could buy cheap German lager in Kwik Save. Yes, it boasted nearly new Ford Sierras and verdigris door knockers. Modern-looking bus shelters and a petrol station with a coffee machine. But Dad wasn't interested in any of that. All he noticed were things that had been taken away. Uprooted hedges, replaced by poorly maintained panel fences. Front gardens that had once been faithfully tended by men called Ray and Cyril, paved over and defended against invasion by ornate iron gates. The wild hedgerows and unkempt edges of fields, cut back and strewn with fag packets, johnnies, plastic bags and drinks cans.

What about the front yard where Dad's mam had welcomed his dad back from Ypres? The row of blackberry brambles that had supplied untold decades' worth of jam jars? The trees he had climbed and the gravel that had grazed his knees? This was a home he no longer recognised. A place he didn't much care for.

'*At least t' sky still lukks t' same,*' I said, as much to myself as to Dad.

'*Aah.*' He nodded.

So that was what we looked at as we set off up Annesley Lane. The low clouds that hung over Felley; the defiant, diffusing arch that bowed to the north west and across the Erewash Valley. In between our admiring glances, we turned left into Portland Road, then onto a grass path that took us to the back of the hill that had once seemed to go on for ever. My flying hill. I hadn't planned to cut across the hill. We just sort of turned into it and were suddenly pushing past familiar leaves and branches. Rolling over thick, fibrous grasses.

And there we were. At the top of the hill, at the top of the estate, at the top of the village. At the top of the whole fucking world. There was a slight breeze, nothing much, but I closed my eyes, tipped my head back and let it skim over the contours of my face. Half a dozen tears dawdled down my cheekbones. Taking their time. The breeze had just enough oomph to lift them, driving them up and back towards Portland Road, over the fields to the M1. I wondered if one had managed to ride its luck, drifting through the air, higher and higher and higher, until it was gently pulled apart above the motorway. Above the hard rumble of the lorries and the howls of the high-tech Japanese motorbikes. Above the irate horns and the speedy flap-flap of partly deflated tyres. Each of its one and a half sextillion teardrop molecules scattered across the East Midlands.

I told my dad about the flying. My flying. The dreams.
I described them to him. I told him that's how I'd sprained
my ankle not long after we moved to the estate. I hadn't
fallen out of my bedroom window . . . I'd jumped.

He laughed, quietly. *'Ah shudd ha' cumm wi' yuh.'*

Maybe the wind heard him. Maybe it was just coincidence.
Maybe. Out of nowhere, a short, sharp gust caught me full
in the chest, knocking me off balance. I let go of the wheelchair and Dad rolled forward a few inches, but his left wheel
was stopped by a partly buried black plimsoll. It was the first
time I'd noticed the patch of ground we were standing on.
Other bits of random rubbish were dotted about on either
side of us: saucepans, coat hangers, smashed ceramics, cushions,
several buckets and a washing-up bowl. The ubiquitous
mattress, of course, bent and leant against the bare frame of
an armchair halfway down the hill.

Another short, sharp gust.

And another, slightly longer and slightly sharper.

All around us, I noticed more and more of the estate's
shit. Torn black bags of soiled nappies and uneaten food.
Mangled bikes and builders' rubble. There was shit, too. Real
shit. Dog shit. Stacked high like sausages in a butcher's window.
And whereas fires had once been a rarity, I could now see
that many of the elderflower trees, bracken and gorse bushes
had been reduced to blackened piles of ash. Even the grass
was silent and sad.

The gusts were becoming regular and much closer together.
Then one long gust, steadily rising in pitch and intensity.
Dad and me watched as it snaked around and in between
the houses on the estate. Careening across the rooftops.
Rattling trees and telly aerials. Our house was in the second
row, just to the left. We watched it for a minute or an hour,
I'm not sure. Long enough for the sky to fill with dark clouds
and dark colours. We saw lights flick on in the chippy and
the pubs and most of the houses on the estate. Not our
house, of course.

I'd always imagined that the wind – my wind, the wind

that carried me to Leicester Forest East – came in a straight line, right down the middle of the estate. As we stood on the hill's brow, I could feel the wind was leaning ever so slightly on my right shoulder. I traced it through the houses, traced its path. Then followed that path all the way back to a tiny segment of the never-ending horizon. There it was, 8 or 9 miles away: Crich Stand. That stately monument to Grandad Jack's regiment, the Sherwood Foresters. Dad must have seen it, too, and slowly adjusted his wheelchair, pulling it back from the partly buried black plimsoll and turning it towards the Stand. Turning it towards his father.

For over 200 years, Dad's family had been buried deep underground, but Grandad Jack and the brave Sherwood Foresters had finally brought them into the light. There they were, standing tall, almost 1,000 feet above sea level. As dizzyingly high as a pit shaft was unnervingly deep. Watching over Nottinghamshire and Derbyshire. Watching over those 200 years. Watching over Dad and me.

And now, for the final time, Grandad Jack was gathering that family together. Great-Great-Great-Grandad Thomas and his sons. Great-Great-Uncle Enoch and his seventeen sons and daughters. Great-Great-Great-Great-Grandad John. All the lost men, women and children.

And now, for the final time, they filled their jet-black, dust-encrusted lungs.

And now, for the final time, they breathed again. A breath so long, so loud, so forceful and joyous that it was racing right across the Peak District, across Amber Valley and the Erewash, across county borders, gravestones, Pit Houses and Clay Heaps. A breath so long, so loud, so forceful and joyous that it would finally lift the last in our family's line – a broken-down man in a wheelchair and a Clever Bugger who wasn't even related to them by blood or bone – clear of that cruel, claustrophobic darkness.

Dad looked up at me and nodded. *'A' tha ready, lad?'*
Yes, I think I was.
Grabbing both handles of the wheelchair, I pushed forward,

then climbed onto the footrests. The hill was much steeper than the road on the estate and we were soon barrelling over the grass and burnt trees, egged on by cheers and hoots of joy that were suddenly ringing out all over Selston. Even Mam was on the front step, happily waving. Unfortunately, she wasn't quite sure where we were and was happily waving at one of Dad's hydrangea bushes.

The wind caught the mood of the crowd and quickly eased me and Dad away from the hillside. Just a few feet at first, enough to clear the trees and hedges, then a long, graceful arch upwards and over the estate. I looked down and saw Dad's wheelchair spinning towards someone's garden on Mansfield Road, landing with a faint thud on the back lawn.

Beneath us, we saw it all. The Parish Hall and Matthew Holland Comprehensive School. The chapels and the Co-ops. The Tin Hat and its car park, full of people who'd all come to see Dad. Pointing and kind laughter.

Dad's mate, Frank, raised a pint of beer towards us. *'Tha'll be oaraight, Norm!'*

A sharp-dressed young miner and his brightly dressed young girlfriend were swinging hips, elbows and knees to 'You Make Me Feel Like Dancing' by Leo Sayer. Sweating and smiling, the young miner showing off his electric-blue three-piece suit and burgundy platform shoes.

'You make me feel like dancin',' he shouted up to me and Dad. *'I wanna dance the night away.'*

One of the Committee men was calling out bingo numbers: *'Eight and one . . .* (dramatic pause) *eighty-one. On its own . . .* (dramatic pause) *number four. Legs-eleven . . .* (dramatic pause) *phweep-phweep.'*

The wind pulled us round to the right and we followed the main road past the Bull & Butcher, Top 'n' Town, down to the Pit Houses and the Rezzer, then back towards Selston Church of England Infant School, Dad's old house at the top of Buxtons Hill, the clean-cut glare of the brand-new White City Estate, the Dumbles, the field with cowpats, The

Shepherds Rest, the Dixies, Bagthorpe Plantation, up the M1, the church, the House of Trees, the green first-aid tin stashed safely in my pyjama top, The Texan, the bustling high streets of Alfreton and Kirkby, the collieries and their colliers.

I turned to look at Dad and wasn't surprised when I saw that his body was no longer bloated and breathless. The fresh air at this high altitude had obviously done him good. He was Pit Houses Dad. Thirteen Trumps on the Staircase Dad. Skegness Dad. He was . . . Dad. Dressed in his favourite green jerkin, worn navy cords and cheap canvas trainers. A brand-new pair, though. Overflowing with sand and coal dust and stars and memories that trailed behind him like the endless vapour from a jet engine.

"Ow long will it tekk t' get t' Skeggy?" Dad asked.

His hand reached out for my hand. I took it. And we laughed . . . like seagulls.

A Final Cough

As mining died and Selston died, my dad died. And his final journey really did begin at the top of that hill, at the top of the estate, at the top of the village. At the top of the whole world. Flying. There was no wheelchair, though. Just me, an urn and a line or two from Zane Grey's *The Last of the Plainsmen*.

The awfulness of sudden death and the glory of heaven stunned me!

Lid off the urn . . . shake-shake-shake. Ashes lifted high by the breeze, then churned and scattered amongst the last scraps of laughing grass and medicinal elderflower.

The thing that had been mystery at twilight, lay clear, pure, open in the rosy hue of dawn.

During the last couple of months of Dad's life, we'd managed a good run of Zane Grey books – I was still friends with Rosemary the librarian and got her to order in as many as she could get. We had even spent one wonderful evening watching Mel Brooks' comedy western, *Blazing Saddles*, on telly. Dad was sitting on the beige and chocolate-brown beast of a sofa, slapping his thigh with joy as he sipped a glass of Manns Brown Ale.

During that same couple of months of Dad's life, Mam's life carried on as normal. She knew something was wrong, but the only hint that she was worried came via her shouting: a bit more and a bit louder than normal. With Dad unable to attempt even the lightest of DIY jobs, she began re-acquainting herself with danger. Attempting to wash the outside kitchen window with the aid of a stepladder, she fell off and broke her ankle. There were less serious moments, too: Mam still talking to a man from the council after he'd gone.

When I pointed out that the chair in the kitchen was empty, she said, '*Ah thought ay were a bit quiet.*'

The summer of 1988 wasn't quite here, but not long after my twenty-third birthday, it sent word of its imminent arrival. That cheered me. The early morning skies and the sunsets. I found a deckchair and Dad would sit out on the back yard, watching me dig the garden, offering bits of advice and telling me where the soil was at its best and where I could store my chitting potatoes. As I was now only at college one day a week, I did as much overtime as possible and bought a proper motorcycle – still thinking of Dennis Hopper and *Easy Rider*, I grew a thick moustache to go with my Honda CX500 Custom. Dad suggested I move the coal pile from the coal shed into the little shed next to the outside lavvy. That way, I could park the motorbike in the coal shed.

On the face of it, that last couple of months was a busy couple of months, but . . . it wasn't. Not really. I spent a lot of my evenings reading to Dad or making sure Mam didn't break another bone. I spent most weekends watching the early morning skies and early evening sunsets. Wondering what was going to happen next.

Then, it happened.

No one had been to see my dad for a while and, to be honest, I don't think he would have welcomed them even if they had stopped by to say hello. He wasn't keen on people seeing his pipe-cleaner, roly-poly body. He spent more and more time in his bedroom, mumbling and sleeping, mumbling and coughing, sleeping and coughing. Coughing and coughing. He'd turned coughing into a language.

When I poked my head around his bedroom door one Saturday morning and suggested that we take another turn in the wheelchair, he looked across at me and shook his head. '*Cooouuuggghhh. Cough-coff. Coff-coff-cooouuuggghhh.*' Delivered with the precise rhythm and cadence of a coherent sentence. So precise and coherent that I immediately understood what he was saying.

'*Worr abaht next wikk then?*' I wondered.

'*Cough-cough-cooouuuggghhh. Coff-coff-cough-coff.*'

So I left it at that.

I stood by Dad's bedroom door and watched him turn onto his side before pulling the heavy woollen sheets and ribbed yellow counterpane over his shoulder. The mumbling and coughing were getting louder and more frequent, so I asked him if he wanted me to call the doctor.

'*Coff-cough-coff. Cough. Coff-coff. COFF!*'

The blue curtains hadn't been drawn and the daylight sat wearily on the odds and ends that took up space in his bedroom. The one picture of his dad; a framed portrait of the Apache leader, Geronimo; a tallboy; a single Zane Grey book on top of the tallboy; a pair of black zip-up Chelsea boots tucked under the bed; Brylcreem, comb and brush on a wooden chair next to the bed. The chair just a few inches from tufts of his grey hair. Poking above the sheets. Flopping and wobbling. In time with the coughing.

I knew that drawing the curtains would only add more weight to the weariness, but it seemed like the right thing to do. It would prevent the world poking its nose in . . . peeping through the bedroom window. Telling everyone about Dad's pipe-cleaner, roly-poly body and the stains on his ribbed yellow counterpane. Why couldn't the afternoon light just leave him alone and let him get on with his coughing and mumbling and sleeping in private? Was that too much to ask?

I watched him fall into a fitful half-doze that eventually settled into something soft and peaceful, then I sat down on the chair. The Zane Grey book on the tallboy was called *The*

Man of the Forest. It wasn't one of the books I'd got from the library. Had Dad got a few others stashed away somewhere? Why hadn't he mentioned them? I flicked through the pages, reading passages at random.

The newcomer entered. He was a large man, wearing a slicker that shone wet in the firelight. His sombrero, pulled well down, shadowed his face, so that the upper half of his features might as well have been masked. He had a black, drooping moustache, and a chin like a rock. A potential force, matured and powerful, seemed to be wrapped in his movements.

I wondered if The Texan had read any Zane Grey books.

Dad's soft and peaceful sleep was so soft and peaceful that he decided to . . . well, he never woke up. Even after I stopped reading and went across the landing to the lavvy, he continued with his soft and peaceful sleep. Even after I went downstairs to watch a bit of *Grandstand*. Even after my mam stomped upstairs and started shouting at him as he lay soft and peaceful underneath the heavy woollen sheets and the stains on his ribbed yellow counterpane.

'Yuh want t' gerrupp an' gerrin t' bath,' she shouted. 'That'll duh yuh gudd.'

Angry shouting. Like me, she was doing what she thought was best.

'C'mon, let's 'ay yuh dahnstairs,' she shouted.

A different kind of shouting. Softer. Frustrated and confused.

'Well, ah'm 'ayin' summat t' eat. Yuh can get y'senn dahnstairs if y' want 'owt. Ah'm non cummin' upp an' dahn all day,' she shouted.

Different again. Hopeful shouting.

Listening to her reminded me of one of those scenes in David Attenborough's wildlife programmes when a penguin or a chimpanzee loses its mate. They know something's wrong, but they continue to prod and pester the lifeless partner. C'mon, stop messing about. Stand up. Have some food. Have a drink of water.

Eventually, my mam stopped shouting, went downstairs and made herself something to eat.

She made Dad a sandwich, too, just in case.

I wore one of my dad's ties at his funeral in Mansfield Crematorium and held my mam tight while she cried. Just like the old days, I kept up a whispered commentary about who'd turned up, where they were sitting, what they were wearing, etc. There was a smattering of Tin Hat mates, ex-mining neighbours . . . a few relatives. Some I knew, vaguely; some I'd never met before. One bloke introduced himself as Dad's brother, my uncle, John. I had no idea that he had a brother called John. Or that he lived in a neighbouring town, just a twenty-minute bus ride from Selston.

I shook his hand and looked at his face, hoping to see my dad's face. Try as I might, I just couldn't think of him as my dad's brother. Going to school with Dad. Sitting down for dinner with him. I should have talked to him – Why didn't you and Dad speak to each other? What was Grandad like? What was Dad like when he was a young lad? – but it's not easy to bond with a relative you don't believe in. Especially at a funeral.

A tall, gangly, bald bloke – I later found out he was my mam's ex-brother-in-law; the 'ex' was because his wife, my mam's half-sister and someone else I'd never met, had died many years ago – did a lot of shuffling and banging as he tried to find a seat. I heard someone else say, *'Ah'm non gooin' t' t' lavvy wi' all them dead folk abaht. It's non raight.'*

A point well made.

I'd chatted to the vicar a few days earlier, trying to pass on a bit of quality info about Dad's sixty-odd years. What the vicar really wanted was all the metaphysical stuff. Thoughts about life, religion, his mates, being a dad and a husband. Sadly, the most I could offer was lapwings and the D500. The jerkin and the Old Holborn tin. Oh, and he'd been a miner for forty years.

There were hymns. For a few minutes, I was back in Selston

Church of England Infant School, belting out 'The Lord's My Shepherd'. Some prayers, a tangential dip into metaphysics, D500s and Old Holborn. Finally, the comforting electric whirr of a motor somewhere behind the velvet curtain.

'Ay's gooin', lukk,' said my mam. 'Ay's gooin'.'

The electric motor sounded smooth and well maintained, but I found myself wondering what would happen if there was some sort of 'fault'. A breakdown. I was now highly skilled in servicing and repairing hydraulic, pneumatic and electrical machines. Would it be right and proper for me to offer up my services? Did the vicar have access to a toolkit and some insulation tape?

I watched as Dad's coffin sank and finally dropped out of sight. No cock-ups from the well-maintained motor.

'Jerusalem' crackled from the small speakers. We weren't supposed to sing this time, but I couldn't help mumbling along.

And did those feet in ancient time, walk upon England's mountains green?

Maybe not the mountains, but he certainly walked upon a few slag heaps and the many green fields by Bagthorpe Brook. Up to the Co-op. Down to the bus stop.

And was Jerusalem builded here, among these dark Satanic mills?

Yes, I think it was.

I wanted Mam to hold the wake at the Tin Hat, but . . . we didn't. So people came back to the house. Uncle John was there. Again, I looked at his face. Was he really my dad's brother? Aunt Lal had made tinned salmon cobs and bowls of crisps. We drank gin and orange (cordial) and Manns Brown Ale. Mam drank too much and laughed too much. Thankfully, there was no sign of Dad's St John Ambulance uniform.

Everybody left and my mam went to bed early. I washed

the pots, opened a couple of Manns and sat in the back room. In Dad's chair. In the reddening early evening twilight. I sat for a long time. I sat and decided to read *The Man of the Forest* out loud to whatever spirits were still lurking in the air.

I was only five or six pages in when my words began to disappear. Drowned out by a whooshed and heavy drumming on the window. A seething rain, blowing hard and fast up the valley from Derbyshire. I pulled back the net curtains and watched as it steamrollered over the estate in careless, uneven layers. It came and kept coming, thrashing the houses and hills and former pits. It rattled the doors of the outside lavvy and the coal shed. It yowled and scratched at the roof tiles, loosing several and flinging them down into the yard. There were times when I heard the crack of branches or the hollow clangour of our dustbin against the side of the house. I saw the back door strain and bend as if a giant shadow was leaning and lurching with all its might. Was the shadow simply trying to escape the storm? Or was it searching for something in our house? Someone? Could it be the shadowy souls of Grandad John, Enoch and Thomas, circling back to pick up what was left of Dad's life?

I heard voices, of course. Whipping around windows, up and down drainpipes, circling my weary, sleeping head. Spinning so fast that their last words were met with new ones. So fast that their new words overtook the old ones. The rain came and it kept coming, until the houses and hills and former pits could take no more.

I woke early. Still in the back room. Still in Dad's chair. Head resting on Zane Grey. The storm had passed, and the morning felt fresh. And warm. In the garden, cabbage white butterflies were bouncing up and down on things that had almost grown. The scent of the scruffy rose bush behind the coal shed had drifted into the kitchen. There were greenfinches hanging from the bird feeders, sparrows nudging past privet leaves and a couple of crows trying their luck on Dad's compost heap.

Well, technically speaking, it was now my compost heap.

I took Zane Grey and a mug of tea – in Dad's mug – out onto the back yard and sat as Dad had sat so many times in his life. On the small strip of concrete that separated the house and the back lawn, knees drawn up, leaning against the house, looking south over the estate. Looking south over Selston.

At first glance, everything seemed as it always had been and as it always would be. Somewhere out there was The Texan and his polished spurs. Grovey and Sadistic-Bus-Stop. The chippy, the Dumbles, the library, Matthew Holland Comprehensive School, the chicken-free Kwik Save, the Tin Hat and the church.

But something had changed. Someone had moved the map . . . turned it upside down. I knew where I was, but all the details had been altered. A house with Venetian blinds instead of curtains. Protective rubber flooring underneath the swings on the Bottom Rec. All those nearly new Ford Sierras and modern-looking bus shelters. Nearly new buses, too. Red ones. The neighbours were people we didn't even know. And hardly any of them went to Skegness.

As a little lad, my rootlessness, though disconcerting at times, had also allowed me to become The Fox, The Shadow, The Mystery Man, a highly trained SAS assassin, the Clever Bugger, the Flying Clever Bugger and the Tin Hat's Dance-Floor Maestro. I had lived a life of excitement, adventure and academic success in a story that had originally marked me down as a nobody.

But now, as a young man, my rootlessness felt heavy and ungainly. Too much for my shoulders. Without Dad, I didn't feel welcome. Stalinistically erased from my own photo album. Suddenly aware that Selston had no need for someone like me. And that worked both ways. What need had I for Selston? *This* Selston. A Selston without mines and miners. Without Pit Houses and clanking coal carts. Without Clay Heaps and a regular clout round the ear. I looked for Pye Hill Colliery. Gone. No pit buildings sending beams of light through the chilly air like golden arrows. Without Dad and without all of that, I had no right to walk amongst those precious memories. I had no right to stand alongside them and call them my own.

I remembered what Dad had said that day we were on the Top Rec: *'There's more on it, burrit feels . . . less.'* I remembered the sad, strained shape of his face as he looked out over a home he no longer recognised.

Now it was my turn.

It must have been the rain. The thrashing rain. Washing every inch of Selston clean as a whistle. Rinsing away the coal dust. Rinsing away that incessant desire to tell common sense to fuck off and mind its own business. Flooding the Dumbles and the gardens and forgotten mine shafts. Pools of water lifting debris from the ground and flushing it into Bagthorpe Brook. A fleck of enamel from an Old Holborn tin, a used airgun pellet, the taste of wild strawberries and nettle beer. Into and underneath the ripples of the brook and then into the River Erewash, steadfastly marking the border between Nottinghamshire and Derbyshire. Between there and here. Between then and now.

I had once wished to wake up on another day in another place. Was this the day and the place? Another place. Another Selston. A pretend Selston.

My Selston reflected in a shattered, ugly, coal-black mirror.

I looked away from the mirror. Up. I could always rely on the sky. Up. That familiar yet ever-changing sky filled with invincible, chugging clouds. Always up. I watched as they listlessly waltzed away to Pinxton and Huthwaite. I listened to the mild breeze, struggling to flutter through the pages of Zane Grey as he sat by my side.

At those moments, it's easy to imagine. To be oh-so sure of other-worldly messages and signs. Were the invincible, chugging clouds a message from Dad? A sign that he was lacing up his cheap canvas trainers and quickly rolling a quick fag? Would that mild breeze peel open the pages of Zane Grey and direct me to a paragraph that would change my life for ever?

A last sentence, perhaps?

A last word?

A last joyous breath?

No. The book was closed, but there would be others. Something dark and funny, maybe. All about a big-headed, badly dressed little lad who fancies himself as the undisputed king of a magical green and grey land.

Oh, and he can fly.